Climate Change and the Management of Natural Systems in Cameroon

Emmanuel Neba Ndenecho

Langaa Research & Publishing CIG
Mankon, Bamenda

Publisher:
Langaa RPCIG
Langaa Research & Publishing Common Initiative Group
P.O. Box 902 Mankon
Bamenda
North West Region
Cameroon
Langaagrp@gmail.com
www.langaa-rpcig.net

Distributed in and outside N. America by African Books Collective
orders@africanbookscollective.com
www.africanbookcollective.com

ISBN: 9956-717-78-9

Dedication

To Hilda with love

Table of Content

Acknowledged Contributions

- Professor Cornelius Lambi, Geography Department, University of Buea, Cameroon.
- Dr. Umar Aliyu Tambuwal, Department of Geography, Usmanu Danfodiyo University, Sokoto, Nigeria.
- Dr. P.A.O. Odjugo, Department of Geography and Regional Planning, University of Benin, Benin City, Nigeria.
- Heinz and Johannes Weigmann Research Award Ecology 2008: acknowledged use of material from published papers authored by E.N. Ndenecho and C.M. Lambi:
    - Climate change and montane biodiversity status in Sub-Saharan Africa: Observations, problems and solutions for the Cameroon Highlands Ecoregion.
    - Climate change impacts on food security and water availability in Sub-Saharan African dry lands: Observations, problems and solutions for the Chad Basin, Cameroon.
- World Wildlife Fund, Washington D.C. for the use of material from the book titled "Buying time: a User's manual for building resistance and resilience to climate change in natural systems".
- Dr. P.M. Oben and Dr. B.O. Oben for the realisation of Chapter Six.
- The use of material from an initial publication, that is, Biological Resource Exploitation in Cameroon.
- My students (past and present) for contributing to the climate change debate issues that motivated me to write this book.

With support from many colleagues.

Preface

As we embrace the new millennium, the threats to nature and biodiversity are unprecedented. Over the years some progress has been made in protecting biodiversity and hydrological systems. These have to some extent proved successful in some areas. However, these efforts are today facing new emerging threats. None of these threats is as great as global warming. The vast array of available scientific literature points to the fact that climate change will have large implications for the way natural resource managers plan and implement conservation strategies.

Traditional natural resource conservation strategies with a focus on confronting human-related stresses on physical and biological systems will flounder in the face of climate change. The problem policy-makers, scientists and resource planners are facing is how to plan and implement strategies for increasing the resistance and resilience of natural systems to climate change. This is because climate change is seemingly difficult to understand and plan for. This book emphasis that planning is essential, as the conservation approaches of the past may not work in an ever-changing warmer environment. The book therefore appraises current management strategies, assesses the biological and physical effects of climate change on natural systems in Cameroon and designs a planning and management framework for each natural system within the context of global warming. Climate change poses a complex bewildering array of problems for ecosystems. The key question is, what can be done – in addition to efforts to reduce CO_2 emissions – to increase the resistance and resilience of these natural systems to climate change?

This book seeks to answer the above question by drawing from the vast array of scientific data available on the subject, and which may not be readily available to policy makers, resource planners, resource managers, environmentalists, students of geography, conservation biology and agronomy. This book constitutes an important manual for those working in Cameroon, that is, those ready to confront the impacts of climate change. It is also a valuable document for teachers involved in the teaching of the functioning and management of natural systems at University level.

Drawing from the expertise of several researchers and experts, the book addresses the major natural systems of Cameroon with practical ideas of how to begin increasing resiliency and of how to plan protected areas in response to the threat of climate change. Some of these strategies can simply be integrated in current management models. Climate change adaptation can therefore be built into current conservation philosophies and plans. It must be noted that, while far from comprehensive, this book brings together assessments and potential initial adaptation strategies for various natural systems represented in the country. This book is a first step in drawing development action to the problem of climate change at local and national levels.

Professor Lambi Cornelius
University of Buea

Chapter One

Global Warming and Development Goals: An Overview

Summary

This chapter is aimed at providing a synthesis of what we currently know about global warming, its possible causes and potential impacts on environment and human activities. The resolutions of the various World Climate Conferences from 1970s to date are also reviewed and presented in this chapter in order to establish how critical the issue of global warming is, and how it dominates the international scene. The current and future challenges of global warming to the successful accomplishment of the development goals of developing countries such as Cameroon are examined. The chapter presents a background to the issues examined in the book. The hope is that stakeholders, planners, resource managers and governments would find the recommendations useful in decision making and policy formulation targeted at building resistance and resilience to climate change.

Key Words: Global warming, development goals, policy-making.

Introduction

The United Nations Framework, Convention on Climate Change (UNFCCC) defines climate change as a change of climate which is attributable directly or indirectly to human activity that alters the composition of the global atmosphere and which is in addition to natural climate variability observed over a comparable time periods (IPCC, 2001). Ayoade (2004) also defines climate change as a long-term shift, alteration or change in the type of climate prevailing over specific location, region or the entire planet. It is obvious from both definitions that change is an inherent attribute of climate, which is caused by both human activities (anthropogenic) and natural factors.

The human factors that cause climate change have been identified as industrialization, technological development, urbanization, deforestation and burning of fossil among others, while the natural factors include solar radiation quality and quantity, astronomical position of the earth. Studies have shown that the currently on-going climate change is caused by anthropogenic factors. Unsustainable industrialization, which releases greenhouse gases, is viewed as the main cause (Odjugo, 1999, 2001; clerk, 2002; Buba, 2004; Nwafor, 2007). Other contributing factors are urbanization, deforestation, burning of fossil fuel and water pollution.

These factors have been observed to alter the climatic conditions of different parts of the world resulting in global warming and devastating extreme weather conditions. The extreme weather conditions include global warming, drought, desertification, flood, sea level rise, wind and rainstorm and thunderstorm among others. Various researchers have studied the causes and effects of the extreme weather conditions as they relate to climate change (Nyelong, 2004; NEST, 2004). On the other hand, there is dearth of information on the economic cost of the impact of climate change. The few known studies include (Odinga, 2004; Buadi and Ahmed 2006; Malun, 2006; Reid, et al 2007) in Mozambique, Cameroon, Cuba and Namibia respectively. Apart from specific countries, other researchers tried to quantify the impact of climate change on natural resources in Africa (Velarde et al, 2005; Macgregor et al, 2006). Interesting enough, none of these studies is based in Cameroon.

Climate change is studied under various facets namely; the science, vulnerability, adaptation and mitigation. The science of climate change has been studied and it is confirmed that the global climate is actually getting warmer (IPCC, 2001, Hengeveld and Whitewood, 2005; Mabo, 2006). The other aspects of climate change have received very little attention especially at regional levels (Mshelia, 2005; Mabo, 2006). This book looks at how vulnerable Cameroon is to the on-going climate change. Vulnerability of climate change refers to the extent to which climate may harm or damage a system. Since this area is among the other facets of climate change that receives little attention, the decision makers may not have much idea about how vulnerable Cameroon is to climate change. So it is the in-depth study of vulnerability of climate change that can reveal the system's sensitivity, and the awareness, understanding and interest of planners, policy and decision makers on the impacts of climate change.

Considering the fact that climate change, global warming in particular, has no boundary, the paper is prefaced by a comment made in 1998 by President Bill Clinton that "Global warming is real; the risks it poses are real;

and the American people have a right to know it and a responsibility to do something about it. The sooner congress understands that, the sooner we can protect our nation and our planet from increased flood, fire, drought and deadly heat waves", President Bill Clinton, July 25, 1998. (Quotation from transcript of President Bill Clinton's radio address of July 25, 1998).

This statement is a mere underscoring of the global warming phenomenon and the danger it poses by confining it to only the American people. This is because global warming affects not only the Americans but all the inhabitants of our dear planet – the earth. Hence, this chapter has attempted to provide a summary of what we need to know about global warming and how its observed and anticipated impacts threaten the accomplishment of socio-economic development and the functioning of natural systems in Cameroon.

First, the consensus of the scientific community has shifted from scepticism to near unanimous acceptance of the evidence of an artificial green house effect. Second, while artificial climate change may have some beneficial effects, the odds, we are not going to like it. Third, reducing emissions of greenhouse gases (GHG) may turn out to be much more practical and affordable than currently assumed.

Although there may not be total agreement on the causes of recent increases in global temperature, the reality of global warming since the mid 1970s is an obvious fact. The question is whether this warming will continue unabated and for how long. This brings us back to the reasons for the current global warming. It is only when we know the causes that we can predict the future pattern or trend of global temperature. Those who believed the major cause of the current global warming is the increase in the emissions of CO_2 and other greenhouse gases have gone ahead to make projection of future trends of global temperature using different emission scenarios especially those of carbon dioxide. They have also gone a step further to make projection of the future climate that would prevail and their possible consequences.

For example, the Fourth Assessment Report (AR4) of the Intergovernmental Panel on Climate Change (IPCC) (IPCC, 2007) confirmed that human actions are changing the earth's climate and creating major disturbances in human systems and ecosystem. The IPCC reports that the world has warmed by an average of 0.76^0C since pre-industrial times. The report also projects that the global average temperature is likely to increase further by 1.8^{0}C to 4^{0}C if no action is taken. The rising global demand for energy, and adverse changes in the climate resulting from effort to meet the rising energy demand, puts the earth and its inhabitants in a "CATCH 22 SITUATION". The challenges confronting us on a global scale are by no means cheap.

In this introductory paper only some of the possible impacts of these climate projections were considered and only an overview account is presented. Those interested in the more scientific details should consult the seminal reports of the Inter-governmental Panel on Climate Change (IPCC) jointly published in 1988 by the World Metrological Organization (WMO) and the United Nations Environment Programme (UNEP). (See amongst others IPCC, 1990a, 1990b and 1990c).

It should be stressed from the outset that global warming in itself if persistent over time will cause a change in global climate. There will be increase in air temperature, rates of evaporation, stronger atmospheric circulation as well as the intensification of the global hydrological cycle. It should be noted that in the introductory part of this introductory paper, it was argued that there were a lot of discrepancies on the recent global warming and for this reason; this paper has also provided a synthesis of the current global warming debate.

The Global Warming Debate

That the earth's temperature is rising is an undisputed fact on which there is virtually no disagreement. There is however some disagreement on two issues namely the causes or cause of the global warming and whether or not the process will continue or will be reversed and how.

It is usual to ascribe warming to the effect of greenhouse gases such as CO_2, Methane, and Nitrous oxides produced merely by human activities. Thus the warming trend from the late 1880s to the mid 1940s has been attributed to the effect of CO_2 produced by industrialization following the industrial revolution of the late 19th century. In contrast, the cooling trend from 1940s to the 1960s was attributed to the cooling effect of aerosols, produced also by industrialization. The global warming since the 1970s has been ascribed to the increasing emission of CO_2, methane, and nitrous oxide by various human activities such as burning of fossil fuel, deforestation, bush burning and other anthropogenic resources (See Table 1).

Table 1: Greenhouse gases and their contributions to global warming

Greenhouse gases (GHG)	Contribution to Global Warming (%)
CO_2	55
Chlorofluorocarbons (CFCs)	24
Methane	15
Nitrous oxide	6

Source: (after Ayoade, 2003)

There is minority opinion in the literature on the debate on global warming that global warming does not exist or that if it exists it has been exaggerated (See Woodcock, 2002). It is argued that though the limited surface network of global weather stations have recorded a rise in the temperature of 0.5^{0}C during the last century including of +0.15^{0}C since 1980, National Oceanic and Atmospheric Administration (NOAA) satellites which measure surface temperature for the whole earth have recorded no rise in global temperature in the last 20 years (i.e. since 1980) for which data are available. It is also argued that readings from available radiosondes match the measurements by NOAA satellites in showing no trend in global temperature. It is then suggested that the surface network of weather stations are perhaps, at least partly measuring urban warming as most of these stations are located in or near urban centres. However, urban warming does not exist in isolation as it helps to raise the general level of air temperature world wide. It has also been pointed out that records from ocean surfaces show a similar pattern of temperature rise to records from land surfaces. Also, data that exclude stations potentially influenced by urbanization also show a rise in temperature.

It is also necessary to note the fact that NOAA satellites do not measure surface temperature; they only estimate temperature of different layers of the atmospheric column and not even the surface temperature. It has also been stated that available rediosounde data, though more limited than the surface records, show a warming of tropospheric temperature of about 0.5^{0}C over the period of 1958 to 1998 in consonance with what the surface data indicate, (Viner, et al, 2000). It is therefore, quite clear from the above, that the fact that the earth's air temperature has been increasing in recent years causing the phenomenon of global warming is well established beyond any reasonable doubt.

When we now come to the issue of explaining why air temperature has been rising the fact of the case is less clear. There are two opposing views on this matter. One says that the warming that has occurred may be due purely to natural forcing. In other words, the greenhouse effect is at work because of increasing emission of greenhouse gases (GHG) through various human activities. The view of the proponents of natural causes of any warming that might have taken place is based on the following premises:

- Warming has been occurring before man-made emissions of greenhouse gases could have had any effects.
- The earth's climate had in the past moved from periods of warm to cold and back again without any man-made or natural causes. For example, the earth was warmer than it is now between 900 and 1300 A.D (the medieval climate optimum) and colder between 1400 and 1800 A.D (the little Ice Age). These large swings in the earth's temperature and others in the past occurred without the impact of man.
- Many solar scientists have shown in their recent research that there is a clear linkage between changes in solar activity and global climate and that solar activity is in fact the dominant factor in global climate dynamics. The solar effect on global climate can be amplified by various factors notably changes in stratospheric ozone and circulatory effects and changes in atmospheric ionization by Galastic Cosmic Rays (GCRs) (Haigh, 2000).

It is quite clear from the above that both variations in solar activity affecting solar output and the greenhouse effect of some anthropogenically produced gases may contribute to global warming but the relative amount of contributions are not exactly known. The problem is that the whole of greenhouse has often been emphasized to the extent that the role of natural forcing due to variations in solar activity has been more or less totally ignored. This accounts for why some people, including scientists have gone to the other extreme of denouncing the greenhouse theory of global warming. Two examples quoted by Woodcock (2000) and one Doerell (1997) will suffice here. Over 1900 people including scientists recently signed Oregun Institute's global warming petition which calls on the U.S.A. government to "reject the Kyoto global warming agreements as there is no convincing scientific evidence that the release of greenhouse gases (GHG) is causing or will cause catastrophic heating of earth's atmosphere and the disruption of the earth's climate". Similarly, over 100 scientists including climatologists have endorsed the Leipzig declaration which states that "most climate specialists now agree that actual observations from weather balloons and satellites show no global warming in direct contradiction to computer climate models". Over 400 scientists including many Nobel Prize winners have signed the 'Heidelberg Appeal' warning the authorities in charge of our planet's destiny against decisions supported by pseudo – scientific arguments or false and irrelevant data (Doerell, 1997).

In similar vein, both the World Energy Council and the International Energy Council and the International Energy Foundation have for obvious reasons opposed the cutting down of carbon emissions. The World energy council has argued that the IPCC recommendations were based on shaky evidence, detrimental to economic growth, unrealistic and influenced by academics seeking to attract funding for their work. On its part, the International Energy Foundation coupled reasons to show that carbon emission is not a threat to the earth's climate. Amongst these reasons are the following:

- Carbon dioxide is not a pollutant but it is vital to life as it stimulates plant growth, increases harvest up to 40% and makes the earth's temperature ideal for life to exist.
- Anthropogenic carbon dioxide is only 4% of all carbon dioxide present in the atmosphere. Reduction of anthropogenic CO_2 will therefore not have any significant impact on the amount of CO_2 in the atmosphere.
- Massive burning of fossil fuels began only 150yrs ago and global climate had gone through cycles of warm and cold periods before then on its own.
- Long-term prediction of climate is unreliable at the time we are not even in a position to predict next week's weather accurately (see Doerell, 1997).

Having examined the different opinions on whether global warming phenomenon is natural or man-induced, the paper now presents a review of some strategies of emission reduction.

Emission Reduction Strategies

The emissions of the greenhouse gases (GHG), particularly the carbon dioxide (CO_2), can be reduced through the following strategies:

- Increase the efficiency of both mobile and fixed sources, in the programme for the Partnership for a New Generation of Vehicles (PNGV), a joint program of the government and the automobile industry which seeks to increase automobile mileage efficiency.
- Increase use of already-proven nuclear energy, a CO_2 emission – free energy source. It already occupies a central role in power production in France and other countries.
- Continue development of new types of energy systems such as fuel cells for use in automobiles and in fixed locations operating on hydrogen stripped from fossil hydrocarbons.
- Carbon sequestration. Another option is to sequester carbon. Many different ways have been proposed:
    - Biospheric sequestration can take the form of growing trees and other plants that use carbon dioxide in the process of photosynthesis. It can be enhanced through biotechnology to produce fast growing trees. Sequestration of carbon in soils is now under consideration.
    - Inject CO_2 into oceans at depth to form CO_2 hydrates.
    - Still other proposals involve fertilizing the oceans by adding iron or phosphorus to increase the production of algae which then would sequester more carbon in the oceans.
- Non CO_2 emissions reduction: These are recent proposals to reduce non CO_2 greenhouse gases such as methane, ozone and nitrous oxide as an effective means of containing greenhouse gas concentration.
- Geo Engineering: There are proposals for geoenforcing involving the dispersion of dust or injection of SO_2 to reduce sun light: These are totally speculative.

Climate Change: A Global Concern

Climate is currently a topical issue worldwide because of the climatic extremes experienced in various parts of the world in recent years. These climatic extremes which have been mainly in form of floods, droughts, heat and cold waves have had a devastating effect on man and his various activities. Climate is not only playing an increasing role in human affairs, there is increasing evidence that man, through his various activities, may influence the pattern of global climate.

Concerned about these effects of climatic extremes and man's interference with global climate which had compounded the current problem of global warming, the nations of the world were forced to rise to the challenges through various forays at both regional and international levels, whereby climate change – related issues were critically examined with the aim of finding lasting solutions to the problems; so that man will continue to live in harmony with his climate. For example, in early 1970s, the entire West African Sahel was hit by catastrophic drought which had caused untold damage to the economy of the region, massive death of people and herd, due to starvation and famine and large scale environmental refugees. Since then, the community of nations had held several international climate conferences to discuss the climate-related challenges and their impacts on man and economy, and possible means of mitigating and adapting to climatic extremes.

For example, the first two world climate conferences have already taken place to discuss issues related to global climate, one in 1979, the other in 1990. Climate issues have featured prominently in global fora such as the 1972 United Nations Conference on the Human Environment, the 1977 United Nations Conference on Desertification and the 1992 United Nations Conference on Environment and Development in Rio de Janeiro, Brazil.

In 1988, the Inter-governmental Panel on Climate Change (IPCC) was established by WMO and UNEP to produce a report giving an assessment of our current scientific understanding of climate change, its environmental and socio-economic impacts and the policy options and strategies needed to respond to such change. Following the excellent work done by the IPCC, the United Nations Framework Convention on Climate Change (UNFCCC) was adopted on 9th May, 1992 at the United Nations Headquarters in New York. The ultimate objective of the convention is to achieve stabilization of greenhouse gases (GHG) concentrations in the atmosphere at a level that would prevent dangerous anthropogenic interference with the climate system and allow ecosystems to adapt naturally to climate change, ensure that food production is not threatened and enable economic development to proceed in a sustainable manner.

In September 1987 the Montreal protocol was signed by 31 countries under the auspices of UNEP to protect the earth's ozone layer by reducing by 50% the production and use of chlorofluorocarbons (CFCs) by the end of the century. In May 1989, the signatories to the Montreal protocol met again in Helsinki along with 30 other countries and agreed to eliminate the production and use of CFCs by the end of the century.

Activities have been taking place at the regional and national levels to understand the mechanisms of climate and the way human activities exert influence on climate. Attention has been given in several countries to the potential impacts of climate change and possible response strategies that can be adopted (see for example, HMSO, 1991; Carter et al, 1994; Ayoade, 1995).

However, the recent and most successful world climate conference so far was the Bali's Conference. About 187 countries met at Bali, Indonesia, December 12 – 15, 2007 under the auspices of the United Nations Framework Convention on Climate Change (UNFCCC) and discussed extensively on various issues of global climate. The UN – Secretary General, Mr. Banki-Moon was reported to have said that the UN breakthrough on climate change had reached in Bali following the unimaginable issues that had been negotiated during the conference. These are: action for adapting to the negative consequences of climate change, such as droughts and floods; ways to reduce greenhouse gas emissions; ways to widely deploy climate-friendly technologies and financing both adaptation and mitigation measures.

The UN Secretary-General further said that there was an emerging consensus on the building blocks of a climate agreement and that a new deal must be comprehensive, involving all nations. He said that developed countries needed to "continue to take the lead on curbing emissions" and stressed the importance of acting globally, whilst providing incentives for countries, business and individuals to act on climate change.

"Our atmosphere can't tell the difference between emissions from an Asian factory, the exhaust from a North American SUV, or deforestation in South America or Africa," he added.

Concluding negotiation in 2009 will ensure that the new deal can enter into force by 2013, following the expiring of the first phase of Kyoto protocol. It can be recalled that early last year, the UN's Inter-governmental Panel on Climate Change (IPCC) issued a finding that if left unchecked, the world's average temperature could rise by as much as 6 degrees centigrade by the end of the century, causing serious harm to environment, societies and ecosystems worldwide.

"This is a real breakthrough, a real opportunity for the international community to successfully fight climate change", said Yvo de Boer, Executive Secretary of the United Nations Framework Convention on Climate Change (UNFCCC). "Parties have recognized the urgency of action on climate change and now provided the political response to what scientists have been telling us is needed", he added. The conference was attended by around eleven thousand participants, among them Secretary General of the United Nations and six heads of state.

Temperature and Precipitation Changes in Cameroon

Ayonghe (2001) showed that the net trend per decade in Cameroon based on data between 1930 and 2000 is a temperature increase of $0.14^{0}C$. The highest rates of temperature increase per decade were at Banyo ($0.24^{0}C$), Maroua ($2.3^{0}C$), Kribi ($0.32^{0}C$) and Bertuoa ($0.21^{0}C$). The lowest rates were at Ngaoundere ($0.06^{0}C$) and Mamfe ($0.04^{0}C$). In the same study plots of total amount of rainfall monitored at 11 stations showed decreasing trends in 6 stations. High rainfall was evident from 1951 to 1967, 1977 to 1980, and 1989 to 1995, while low rainfall was observed from 1930 to 1950, 1968 to 1976, and 1981 to 1988. In the drier regions of North Cameroon the data between 1960 and 1990 were in agreement with similar studies in the north-eastern arid zone of Nigeria where Carter and Alkali (1996) reported rainfall as decreasing by 8mm per year. The net changes in temperature across the entire country from 1930 to 1935 were $0.91^{0}C$, which is equivalent to $0.14^{0}C$ per decade. Similarly, the net change in the total amount of rainfall was minus 282mm, equivalent to minus 43mm per decade. The net change in the annual number of rainy days was minus 7 days over the same period. When projected to the year 2060, these net changes give a temperature increase of $1.8^{0}C$ and rainfall decrease of 559mm, and 16 days respectively. The changes in mean number of rainy days are presented in table 2.

Table 2: Mean total number of rainy days computed from trend lines.

Stations	1940	1960	1980	1995	2030	2060
Garoua	82.1	79.7	77.2	75.4	71.1	67.4
Ngaoundere	145.9	146.8	147.7	148.4	150.0	151.3
Mamfe	248.4	222.4	196.5	177.0	131.5	92.6
Bamenda	213.5	204.3	195.0	188.0	171.8	157.9
Banyo	138.4	151.6	164.9	174.8	198.0	217.8
Maroua	72.8	71.8	70.8	70.1	68.4	66.9
Douala	142.8	229.6	216.3	206.4	183.3	163.4
Kribi	209.0	205.1	201.1	198.2	191.3	185.4
Yaounde	148.9	149.1	150.5	151.2	152.9	154.3
Ebolowa	156.4	163.8	171.2	176.8	189.7	200.9
Bertoua	123.1	127.4	131.8	135.0	142.6	149.1
Mean	**161.9**	**159.1**	**156.6**	**154.7**	**150.0**	**146.1**

Source: Ayonghe, 2001

The effects of global warming on air and water temperatures are likely to be far more complicated than a gradual increase in average air temperatures. Both observational studies and models of future climate change suggest that there will be more variable temperatures, a greater frequency of extreme temperature events and more hot dry season days. They also predict a change in rainfall variability and an increase in the frequency of extreme rainfall due to an increase in the intensity of tropical storms. These will result in increased droughts and floods.

The weather and climate institutions of Cameroon, research institutes and universities should be updated by installing more conventional and upper air stations in the country so that their capacity for generating climatic data will be enhanced. This is necessary for climate projections and the sustainable management of natural systems. A National Climate Programme (NCP) should be launched in Cameroon geared towards human and institutional capacity building in climate change studies. This should be affiliated to similar programmes worldwide. Such institutes will generate climate data that will fashion government policy in the management of natural systems.

Impact of Global Warming on Development Goals

Agriculture and Land Use

Agricultural production and land use may be affected in diverse ways by global warming and climate change. The increase in CO_2 content of the atmosphere largely considered to be the primary cause of global warming will have impact on agricultural production. Similarly, the increase in air temperature and consequent increase in rates of evapotranspiration will also affect agriculture directly and indirectly. Climate change may involve a shift in climatic zones or rain belts especially in the tropics. The effects of such shifts in rain belt on agriculture may be desirable or undesirable depending on whether or not the envisaged increase or decrease in precipitation amount favour crops.

Global warming is expected to cause increase in climatic extremes and more variability in climate. There would be increase in the risk of crop failure as a result of higher frequencies of drought, flood, storms and other weather hazards to agriculture. A rise in sea level that would result from continued global warning will also affect agriculture adversely especially in coastal and deltaic areas. The outcomes of these effects are the sharp reduction in crop yield which will lead to hunger and starvation. So, the aim of halving extreme poverty and hunger which is the first of the eight goals of the Millennium Development Goals (MDGs) is at stake if global warming should continue unabated.

Natural Systems

Similarly, the increases in atmospheric concentrations of CO_2 and other green house gases and the resultant global warming and projected changes in climate will have impact on *natural terrestrial ecosystem* in diverse ways. Changes in temperature, precipitation and relative humidity will cause ecological stresses in marginal areas and will affect the dynamics and distribution of plant species and animals including insects. Therefore, the aim of ensuring environmental sustainability, the seventh of the Millennium Development Goals (MDGs) may not be realized if global warming should continue unchecked.

Human Settlements

Equally, *human settlements* will be affected directly and indirectly by global warming and climatic change. The most vulnerable human settlements will be those that would be affected by climate change induced natural hazards such as coastal and river flooding, severe rain storms, tropical cyclones and heat waves amongst others. Coastal human settlements will be adversely affected by global warming induced rise in sea level. Human settlements in small island nations will be similarly affected. Many coastal areas will be inundated as the sea level rises while many such areas in the tropics and sub-tropics will be affected with additional threat of damage by more intense tropical cyclones spawned by increases in sea surface temperature as a result of global warming. Therefore, all plans to ensure environmental sustainability may not materialize if emissions and concentration of greenhouse gases (GHG) in our atmosphere continued unabated.

Human Health and Disease

In addition, global warming and climate change will have direct and indirect impacts on *human health and prevalence rates of diseases* of various types. The increase in temperature will cause an increase in the incidence of health disorders due to heat especially in the temperate zone where the increase in temperature especially in summer will be highest. The urban areas already known for their heat island phenomenon will be most affected. Heat related disorders include skin rashes, prickly heat, heat exhaustion and heat stroke. Salt depletion which occurs under hot conditions often manifests in cramps, fatigue and anorexia.

Other potential impacts of global warming and climate change on human health are indirect. For example, food shortage which was one of the resultant effects of global warming on agriculture could have adverse effects on nutritional status of people especially the poor ones with implications for their health, productivity and well-being.

The Way Forward

This chapter will be concluded by making some suggestions on the way forward in respect of coping with both the current and future challenges of global warming and associated climate change, so as to ensure the successful accomplishment of Development Goals. However, a reference has been made to the acceptance speech of the former United States (U.S) Vice President, Al Gore, after the acceptance of the Nobel Peace prize in Oslow, Norway, Monday, December 10, 2007. The speech was titled: "We are confronting a planetary emergency". Al Gore won the Nobel Peace Prize together with the team of scientists of Intergovernmental Panel on Climate Change (IPCC) for his tremendous effort in promoting international awareness of the global warming phenomenon and its associated challenges to both our dear planet the earth and its inhabitants.

So, a synthesis of what we currently know about global warming and its challenges will not be comprehensive without making suggestions on the way forward as aimed by this chapter. The paper therefore suggests that, while the climate change mitigation strategies should be adopted, the climate change adaptation strategies should also be formulated.

Mitigation strategies are measures taken to deal with causes of climate change. They are actions taken by man to prevent or retard the increase of greenhouse gases (GHG) concentration in the atmosphere. This may be achieved by limiting current and future emissions from man-made sources of greenhouse gases (GHG) by enhancing the potential sinks of greenhouse gases.

Climate change *adaptation* refers to the responses to both the adverse and positive effects of climate change. Adaptation is any adjustment made whether passive, reactive, or anticipatory in response to anticipated or actual consequences of climate change.

The paper therefore suggests six types of strategies for adapting to the effects of climatic changes as identified and recommended by Carter *et al* (1994).

These are:

- **Prevention of Loss**: This involves anticipatory actions or measures taken to reduce the susceptibility of an exposure unit to the impacts of climate change. An exposure unit is defined as the activity, group, region or resource exposed to the effects of climate change.
- **Tolerating Loss**: Losses may be tolerated where adverse impacts can be accepted in short-term because they may be absorbed by the exposure unit without long term damage. In other words, nothing is done to protect the exposure unit against the effects of climate change.

- **Spreading or sharing Loss**: Here actions are taken to redistribute the burden of the impacts of climate change over a larger region or population beyond those directly affected.
- **Changing Use or activity**: This involves a change in activity or source use to adjust to the adverse as well as the positive effects of climate change. For example, there could be a switch from cultivation of water-demanding crops to crops that are less water demanding. For example, millet may by grown instead of guinea corn or maize.
- **Changing location**: An activity is relocated to a more suitable location under the changed climate. For example a hydro-electric facility may be relocated due to a change in water availability. Also, a settlement of industrial plant may be relocated to avoid inundation by a rise in sea level as a result of climate change.
- **Restoration**: Here, the aim is to restore an exposure unit to its original state following damage or modification as result of climate change. Strictly speaking, restoration is not adaptation to climatic change since the exposure unit is still susceptible to subsequent comparable climatic events. A good example is the restoration of a building or an historical monument that is susceptible to flood damage.

The available literature now makes it abundantly clear that the climate is changing and ecosystems are being affected by global warming. There is already much awareness on the anthropogenic impacts on natural systems. Conservation practitioners and policy-makers must now address climate change. This book seeks to assist them to consider how to respond to this growing threat.

In each of the case studies discussed in this book there are examples of specific climate-related changes. Clearly most natural systems will be dramatically challenged and subsequently altered by climate change. It is therefore unlikely that any current local management and conservation strategies could provide adequate protection for biodiversity under climate change conditions (see Table 3).

Table 3: The effects of climate change on biodiversity for Three Thermal thresholds (adapted from WWF, 2003).

Temperature change (^{0}C)	Effects on biodiversity
2^{0}C	Some species lost. Possible management options exist
4^{0}C	Many species lost Few management options. Those that exist will be extremely expensive
6^{0}C	Dire

From the above it can be seen that conserving biodiversity will therefore require a two pronged approach:

- First, greenhouse gas emissions must be dramatically reduced in order to slow the rate and extent of global climate change.
- Second, assuming that we can limit the rate and extent of change, we will still need to respond to the change that is already inherent in the system and buy some time for ecosystems as emissions are reduced. This is because these gases in the atmosphere have a time lag.

Conclusion

We must therefore take local action to protect biodiversity by increasing the resistance and resilience of natural systems so they can better survive the changes to come. There is the need to integrate climate change threats into conservation plans. Conservation planning is the key to protection of biodiversity and ecosystem function. The majority of planning to date has focused on issues relating to space; designing reserves and enacting protection laws. Protection to date has focused on anthropogenic impacts. Emerging environmental threats such as climate change require that governments extend conservation planning beyond the boundaries of protected areas, and into a future in which ecosystems and biomes may be quite different than they are now. While it is incumbent on us and governments to take action now to design, test and adopt conservation strategies that respond to climate change, these efforts are not the long-term solution. These strategies to increase resistance and resilience of ecosystems will work only for changes of a few degrees (^{0}C) (see table 3) at most. In essence, we are only buying time, but time that natural systems desperately need while efforts are made to stabilize atmospheric concentrations of greenhouse gases and limit the rate and extent of climate change.

Acknowledgements

The author acknowledges with thanks Dr. Umar Aliyu Tambuwal and Dr. P. A. O. Odjugo whose respective papers were used to realized this overview:

- Global warming and the millennium development goals in Nigeria: the way forward;
- Quantifying the cost of climate change impact in Nigeria: emphasis on wind and rainstorms; all in: Book of Proceedings: *Geography and the Millennium development goals: translating vision into reality.* 50th Annual Conference of the Association of Nigerian Geographers, University of Calabar, Nigeria.

References

Akande, A. (2008) "*Millennium Development Goals in Nigeria*", retrieved April 18, 2008 from http://www.one world.net/Guides/Nigeria/development.

Ayoade, J. O. (2003) Climate change. Vantage Publishers Ltd, Ibadan.

Ayoade, J. O. (1995) Climate and human welfare. Inaugural lecture delivered at the University of Ibadan, Thursday, June 15, 1995. Ibadan University Press.

Ayoade, J. O. (2004) Climate change. Ibadan. Vantage Publishers. Pp 45 – 66.

Ayonghe, S. (2001) A quantitative evaluation of global warming and precipitation in Cameroon from 1930 to 1995 and projections to 2060: effects on the environment and water resources. In: E. B. Eze and C. M. Lambi (eds). *Readings in Geography.* Unique publishers, Bamenda p. 142 – 155

Buadi, P. and Ahmed A. (2006) The impacts of wind related hazards in Southern Cameroon. *Journal of Physical Sciences.* 3(2): 82 – 88.

Buba, A. (2004) Climate change and water problems in Chad Republic. *Journal of Arid Environment.* 3(2): 42 – 50.

Carter, T. R. et al. (1994) *IPCC Technical guideline for assessing climate change impact and adaptation,* UCL, London.

Carter, R. and Alkali, A. (1996) Shallow groundwater in the North eastern arid zone of Nigeria. *Quarterly Journal of Engineering Geology*, Vol. 29, p. 341 – 355

Clerk, C. (2002) Climate change or climate fluctuation? *Journal of Arid Environment.* 1(1): 18 – 33.

Clinton, B. (1998) "*Threat of climate change*". Quotation from transcript of President Clinton's Radio Address of July 25, 1998. Retrieved April 28, 2008 from: http://www.ecobridge.org/content/tht.htm.

Gore, A. (2007, December 10) "We are confronting a planetary emergency". *Focus*, p. 13

Haigh, J. D. (2000) Solar variability and climate, *Weather, Vol. 55*, No. 11, pp. 399

Hengeveld, H. and Whitewood, B. (2005) Understanding climate change 2005: A synthesis of climate change science. *Environment Canada*, p. 31 – 35

IPCC (2001) *Climate Change 2001 Third Assessment Report*, Vol. 4

IPPC (1990a) *Climate change: The IPCC Scientific Assessment.* Cambridge University Press, Cambridge.

IPCC (1990b) *Climate change: The IPCC Impacts Assessment.* Australian Government Publishing Services, Cambera.

IPCC (1990c) Climate change*: The IPCC response strategies.* WMO/UNEP.

IPCC (2007) Climate change 2007: Fourth Assessment Report. Cambridge, Cambridge University Press.

Mabo, C. B. (2006) Temperature Variation in northern Nigeria between 1970 and 2000. *Journal of energy and environment* 8(2): 44 – 60

Malun, C. (2006) Socio-economic impacts of climate change in Cuba. *Journal of Environmental Science* 8(2): 44 – 60

MacGregor, J; Reid, H; Sahlen L. (2006) An agenda for change: Quantifying climate change impacts on natural resource-based economies. *Sustainable Development Option Papers. October, 11057IIED*

Mshelia A. (2005) Adaptation strategies to climate change. *Journal of Energy and Environment.* 18(3): 74 – 81

Nigerian Environmental Study/action Team (NEST) (2003) Climate change in Nigeria: A communication guide for reporters and educators, NEST. Ibadan.

Nwafor, J. (2007) Global climate change: The driver of multiple causes of flood intensity in Sub-Saharan Africa. Paper presented at the International Conference on Climate Change and Economic Sustainability held at Nnamdi Azikiwe University, Enugu, Nigeria. 12 – 14 June 2007.

Nyelong, P. (2004) Global warming and global waters. *Journal of Energy and Environment.* 17(1) pp. 79 – 90

Odinga, S. (2004) Evaluation of the economic cost of climate change in Mozambique. *Environmental Impact Analysis* 12(1): 45 – 52

Odjugo, P. (1999) Climatological implications of environmental degradation on sustainable human development. *GeoResearch.* 3: 52 – 60

Odjugo, P. (2000) The impacts of global warming on climate change and extreme weather conditions: Global and regional evidences. *Asia Pacific Journal on Environment and Development, Vol. 7*, p. 53 – 70

Reid, H: MacGregor, J; Sahlen, L. and Stage (2007) Counting the cost of climate change in Namibia. *Sustainable Development Option Papers.* December, 17026IIED P2. http://www.iied.org/pubs/display.php = 17026IIED. Assessed 13/5/2008.

STN (2008, April 28) Global warming and the energy test: The African position. A research work by the STN publication group. *Daily Independent*, p. EM 3 – 4.

Velarde, S. j., Malhi, Y., Moran, D., Wright, J. A. and Hussain, S. (2005) Valuing the impact of climate change on protected areas in Africa *Ecological Economics*, 53, (1), 21 – 33.

W. H. O. (1990) Potential Health Effects of Climatic Change. W. H. O., Geneva.

Woodcock, A. (2000) Global warming: The debate heats up. *Weather*, 15, p. 143 – 144.

UNFCCC (2007) Proceedings from the world climate conference in Bali, Indonesia, December, 12 – 15, 2007. Retrieved April 14, 2008 from http://unfccc.int/meetings/cop_13/items/4049.php

Chapter Two

Climate Change and Rainforest Ecosystem Management

Summary

The future of the rainforests of the world is high on the agenda of international organization agencies. This is because of the value of their biological resources, that is, consumptive use value, productive use value and the benefits occurring to society as ecosystem services. The size of the rainforest in Cameroon, the great biological diversity, and the current anthropogenic pressures on the ecosystem call for urgent conservation attention. There is need to evaluate how resilient the forest will be in future once an overlay of probable climate change impacts set in, that is, a detailed level of analysis of vulnerability to climate change. The chapter analyses these threats in combination, appraises current forest management strategies, and identifies the scope for selecting adaptation options and management strategies that could promote resistance and resilience to climate change. It concludes that the selection of adaptation strategies should depend largely upon the existing and eminent non-climatic stresses to the forest system. This is the baseline upon which climate change will exert additional pressure. The overlay of climate change-induced threats pointing to particular species or system vulnerabilities will aid in the choice of strategies. However, the reduction of greenhouse gas emissions remains the only long-term option to ensuring healthy forest ecosystem.

Key Words: Climate change, rainforest, resistance, resilience, human-induced pressures, vulnerability analysis, adaptation options.

Introduction

Tropical forests cover 6% of the earth's land area and are found in the equatorial zone of Latin America, Africa and Asia. The tropical rainforest, as we now call it, seemed immense and almost indestructible. It first began to disappear in the Caribbean islands as early as 1671 after sugar cane plantations swallowed up forest on Barbados. Other islands in the region went through similar experience, a foretaste of a global trend that has accelerated in the 20th Century (Anon, 1998).

Today tropical rainforests carpet only 5% of the earth's surface, compared with 12% a century ago. Every year an area of about 130.000 km^2 of forest is felled or burned and about 56% the world's tropical forests have been cleared or damaged (Miller Jr., 1994). This appalling rate of destruction threatens to condemn the rainforest and its ecosystems. Forests in the developing countries are in a crisis and facing serious degradation. In 1980, there were about 646 million hectares of forests and wood lands in sub-Saharan Africa. A 1980, FAO / UNEP study estimated 4.7 million hectares of these forests were being felled annually (Lanly, 1982).

The Cameroon rainforest has an approximate area of 200,000 km^2. This forest harbours 297 species of mammals, 949 species of birds and 3000 species of amphibians (Ngome, 1992) and 156 endemic species (Gartland, 1992). The size of the forest and the great biological diversity calls for urgent protection efforts. On the other hand, the contribution of the forestry sector to the national economy is significant. Timber alone accounted for 4% of GNP in 1991, generated 32 billion CFA francs worth of exports and provided 20.000 jobs (MINEF, 1992). (Figure 1 and 2)

The future of the rainforest of the world is high on the agenda of international organizational agencies at present (Rio, 1992). Cameroon has one of the largest reserves of rainforest left in the world which extends into neighbouring Congo, Gabon, and Democratic Republic of Congo. A debate exists between the major conservation agencies who want to protect these forest and commercial / state interests who exploit the forest for timber and natural resources.

The chapter appraises forest management strategies in Cameroon with a focus on sustainable management, climate change impacts, identifies the scope for selecting adaptation strategies and a policy framework for the mitigation of impacts.

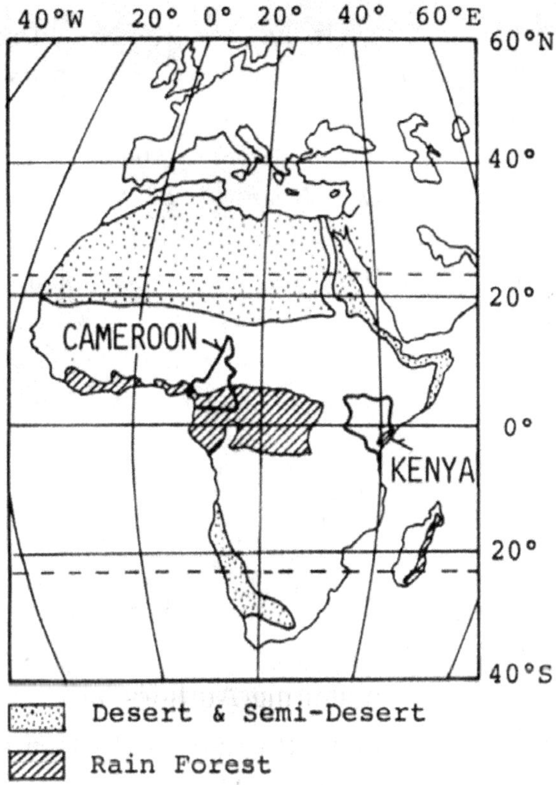

Figure 1: *Location of the Cameroon and rainforest in Africa*

The Extent of Forest Exploitation

- Timber Reserves: 7 million hectares with an estimated potential of 29,400,000m3 of *Entandrophragma* and 2.4 million m^3 of *Khaya.* These are undergoing development with the objective of timber production.
- Exploitation Zones: 8 million hectares opened to exploitation. Standing timber is estimated at 24 million m^3 of *Entandrophragma* and 1.96 million m^3 of Mahogany. It is estimated that only 250.000 hectares of these forests are exploited yearly and less than 10 cubic metres of timber are harvested per hectare. This is in order to minimize the resulting disturbances suffered by the forest from extraction, paths or tracks and camps. These affect according to local

forestry experts are supposed to be only 5% of the exploited area. This is however not the case.

The inventories of major marketable species were made in 1982 on 14 million hectares of rainforest area.

- *Entandrophragma cylindricum* (Sapel) : 42,032,400m^3
- *Entandrophragma candollei (*Tiama): 1,759,300 m^3
- *Entandrophragma* utile *(*Sipo): 4,684,300 m^3
- *Entandrophragma congolensis (*Kossipo): 8,535,900 m^3
- *Khaya Sp.* (Mahogany) 8,087,600 m^2
- *Afromasia Spp.* (Afromasia): 8,087,600 m^3

The development of the forestry potential involves afforestation. Most forest plantations were established in the 1940s. The main species planted include: *Melianceae* (Bibolo, Sapeh, Sipo), *Combretaceae (*Framine, Frake), *Sterculiaceae* (Doussie, Assamela), *Mosaceae* (Iroko), *Myristiacrae* (Iloba), *Sapotaceae* (Moabi), *Ochnaceae* (Azobe) and *Burseraceae* (Koume). Before 1982 an approximate 16,306 hectares of rainforest were planted. Between 1982 and 1990 about 5,905 hectares were planted. Sustainability is facilitated by:

- Rapid regeneration: Abundant fruits, and easy germination due to the abundance of seeds and propagation possibilities.
- High sunshine demand: The seedlings require more and more light as they grow in order to be fully exposed to sunlight at the adult stage. *Entandrophragma* exploitation is therefore important and necessary for the survival of the species.
- Wide ecological range: As concerns Cameroon, according to officials the above species are cosmopolitan and much of the forest potential is still intact.

Fourteen million hectares of rainforest in the country have an exploitable potential of 1.517 billion cubic metres (all species), 562 million cubic metres are presently being marketed and *Entandrophragma* alone contributing 57,011,900 cubic metres of timber to the marketed stock (Table 1).

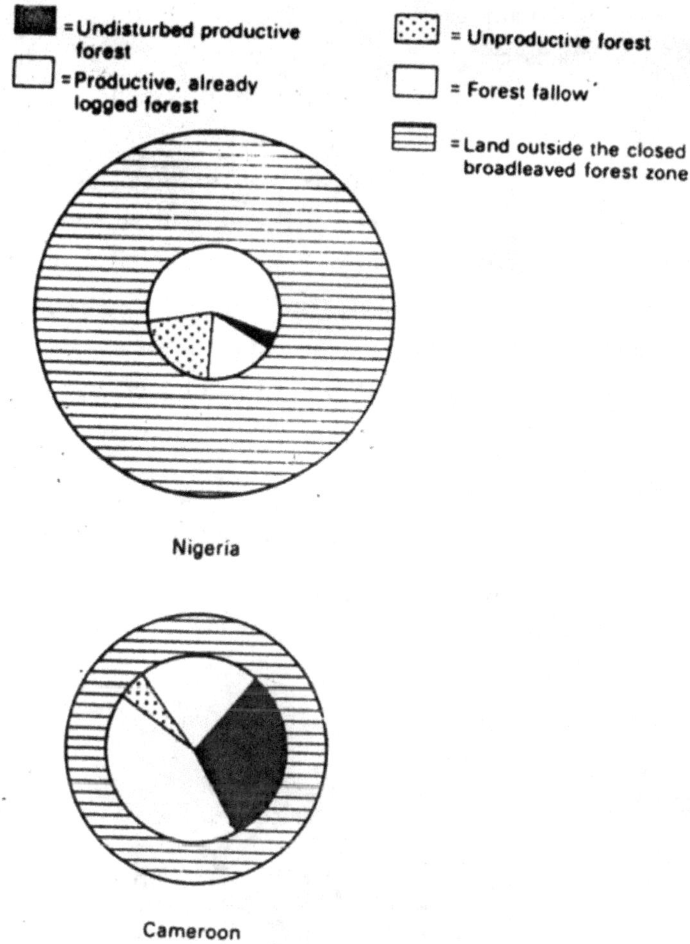

Figure 2: *The extent of conservation in the area of closed and broad-leaved forest in Cameroon and Nigeria.*

Table 1: *Annual average exploitation of marketed timber species*

Species exploited between 1983 and 1993	Annual production (m^3)
Entandrophragm angolaise (Tians)	2,226
Entandrophragm condolei (Kossipo)	17,865
Entandrophragm utile	36,903
Entandrophragm cylindricum	296,935
Khaya anthotea / ivorensis (Mahogany)	22,431
Pericopsis elata (Afromosia)	5,134
Total	**381,544**

Source: Cameroon Union of World Producers and Exporters.

A development path is sustainable if and only if the stock of overall capital assets remains constant or rises over time. The asset in question is environmental capital, that is, the rainforest. To be on a sustainable development path, then, a nation must be living within its means, which, in this context, means not decreasing its overall capital assets. The proper measure of income corresponding to this idea of sustainability is widely accepted to be the amount that can be consumed without running the stock of capital down. In order to measure the viability or sustainability of these species at the present harvesting rate, we can establish a relationship between the available volume of timber and the average volume harvested yearly.

Table 2 gives an idea of the sustainability of these species if the present harvesting rate is maintained and if no sustainable efforts are made. Of course, this is not likely going to be the case. This is because of current government investments in developing forest resources, promoting preservation and sustainable harvesting. Forestry experts in Cameroon argue that these species cannot be extinct and are not in danger because of their specific reproductive characteristics, their cosmopolitan nature and the low human population density of 3 to 4 inhabitants/km^2 in the rainforest zone.

Photo: *This forest road is built to evacuate timber from the rainforest. It encourages further settlement and destruction of the primary forest*

The government therefore recognized the rainforest as a national patrimony, where in the state acts as either the trustee for, or the actual owner of all natural resource endowments. The government holds property rights over the forest. The government plays an active role in the forest resource development by functioning not only as tax collector, but as seller or leaser of exploitation rights, gate keeper for the entry of foreign firms, and often as a business partner with private domestic and foreign firms.

Table 2: *Determination of the exhaustion time of species at present harvesting rate*

Species	Available potential 60% of dense forest in m³	Annual volume exploited in m³	Estimated time of exhaustion (years)
Entandrophragm angolaise (Tians)	1,759,300	2,226	760
Entandrophragm condolei (Kossipo)	8,535,900	17,865	477
Entandrophragm (Sipo)	4,684,300	36,903	127
Entandrophragm cylindricum	42,032,40 0	296,935	141
Total Entandrophragm	**57,011,90 0**	**381,544**	161
Khaya (Mahogany)	3,036,300	22,431	135
Pericopsis elata (Afromasia)	8,087,600	5,184	1560
Total of all species	**68,135,80 0**	**381,544**	178

Source: ONADEF Reports

Although Cameroon has realized substantial increase in GDP during the last decades as a result of timber exports it is questionable how long this will continue. The bad news is that deforestation rates have been high and are rising. IUCN/UNEP (1986) estimates Cameroon has a closed forest area of 17,920,000 hectares and an annual percentage deforestation rate of 0.4%. The original wildlife habitat was 46,940,000 hectares, the amount remaining by 1981 was 19,245,000 hectares making a habitat loss of 59%. Timber in general is viewed as a renewable resource. This is clearly evident for the temperate forests of Germany and the vast stands of pines found all over North America. Tropical hardwood are also renewable in principle; left alone a cut-over tract will ultimately yield new adult stems, although not necessary of the same species extracted. But although tropical timber may be renewable, it is not all together clear that tropical forests are renewable. Several factors have made many forests specialists to classify this resource as "semi-renewable." The factors include:

- The long growing cycle (40 to 150 years) for most species of logs harvested;
- The fragile ecology of the tropical forest; and
- Except for the teak, the widespread lack of success in inducing regeneration of harvested species.

Anthrogenic Degradation of Forest Environments

The rapid consumption of tropical timber reserves was promoted by the liberal development policy of the country. Raising export production was given top priority. Export duties constituted one of the states most important sources of revenue. The relatively high revenues from the timber exploitation and wood processing industries, however, were seldom coupled with a reflux of funds. Only a small proportion of tax revenues is utilized for forestry management. The work of the forestry authorities is accordingly ineffective. Examples of this are insufficient monitoring of exploitation concessions (Figure 3). The forestry staff was, and is today, forced to earn additional money through corruption, since salaries are not enough to sustain the usually large extensive African families. Trade in timber and the development of timber exploitation has contributed to the destruction of forest reserves.

First through direct interventions in the ecosystem in the form of cutting and removal of timber. As a rule, only a few trees per hectare prove more harmful to the ecosystem than it is generally assumed. The felling of the trees above causes considerable damage. In general, lianas are not cut before felling, so that numerous other trees in the vicinity or their crowns are pulled down as well. Furthermore, the falling trees destroy a large part of the natural regeneration. Even more damage can be attributed to the use of heavy machinery and the building of forest roads and haulage ways. Without increasing follow-up measures, this has resulted to a diminishing of the regenerative capacity of the forests and precious wood varieties. While there is an initial increase in overall regeneration, it occurs mainly among the usually not very useful pioneer tree species and the less desirable timber stands. The more valuable tree species lose their capacity to regenerate through removal of large-diameter trees (Jurgen, 1990).

Photo: *Loggers and roads destroy the rainforest – making it susceptible to invasive species and climate change induced vegetational changes*

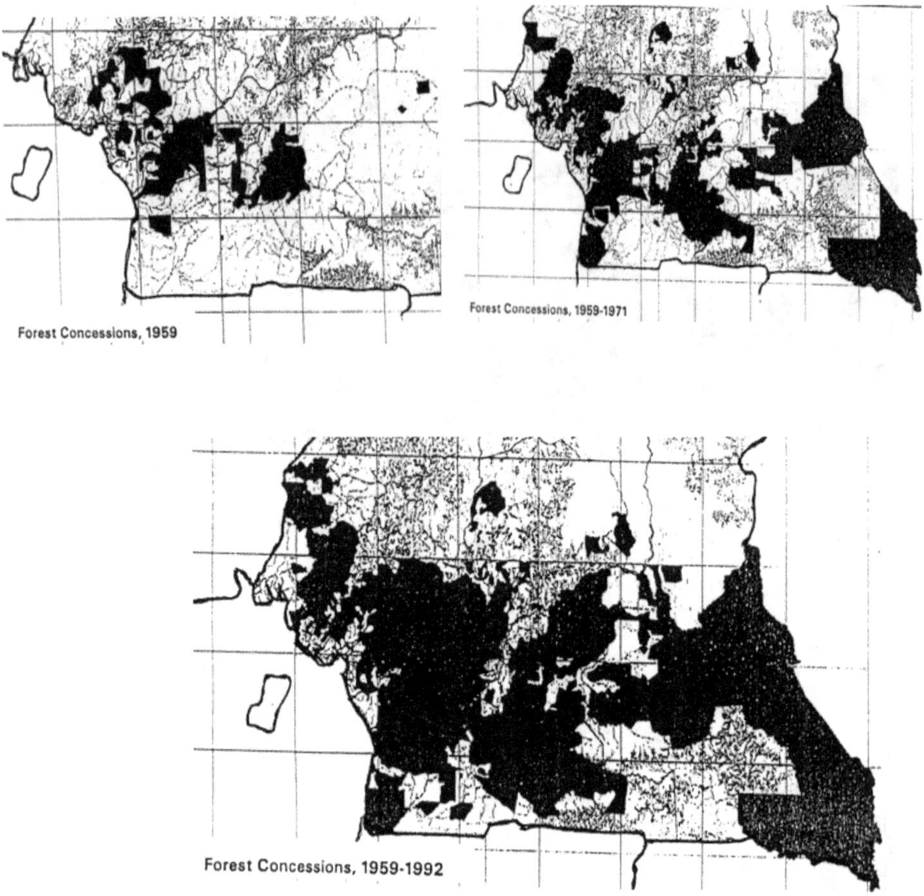

Figure 3a: *Allocation of forest concessions in Cameroon*

The opening and clearing of forests is a starting point for agricultural use. The exploitation of tropical wood has paved the way for a speedy transformation of forested area into agricultural land. The timber exploitation front penetrates the forest deeper and deeper followed by subsequent colonization by agricultural land. The first to settle in the cleared areas are tree harvesters who through their work with the timber companies created a capital base for the purchase of tools and food. These are subsequently followed by a stream of settlers from other regions. The environmental consequences of this human impact on the rainforest are presented in figure 4.

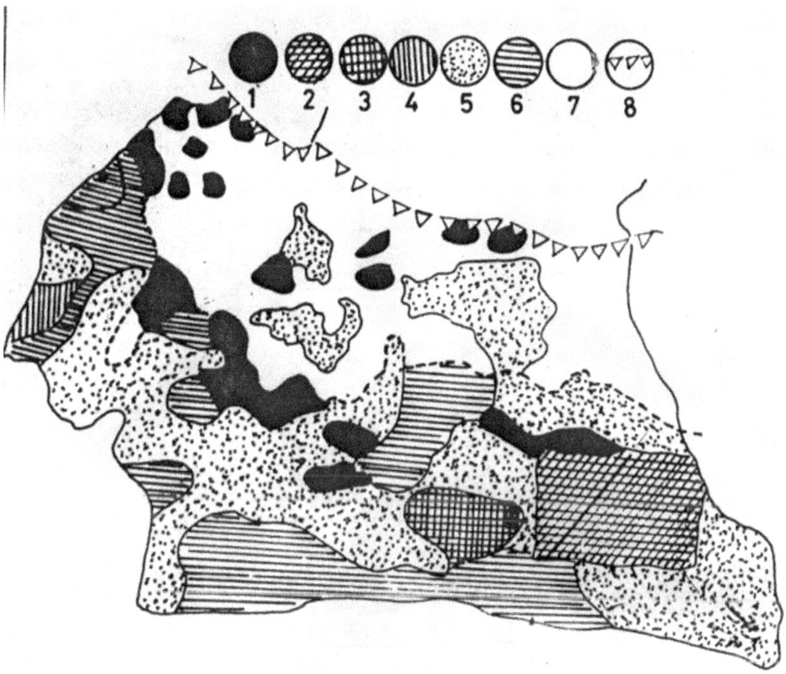

Figure 3b: *Shifting forest boundaries in Cameroon: a slow desertification process: - 1 = forest not greatly degraded, 2 = Yokodouma Reserve, 3 = Dja Reserve, 4 = Korup Reserve, 5 = Intensively degraded forest, 6 = Forest degraded by industrial exploitation, 7 = High Savanna woodland derived from degraded rainforest, 8 = Possible original northern limit of the rainforest.*

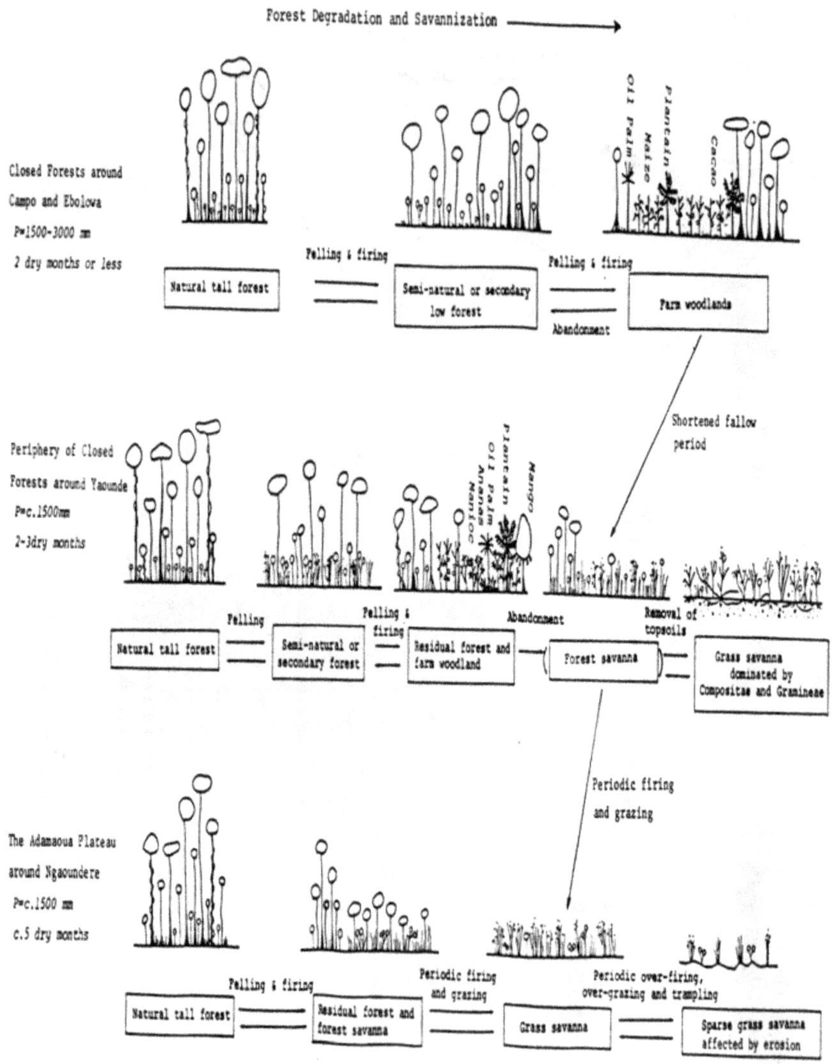

Figure 4a: *Dynamic relationship between human impact and vegetation, and processes of savannization in the humid areas of Cameroon (modified from Haruki, 1984). See Figures 4b and 4c.*

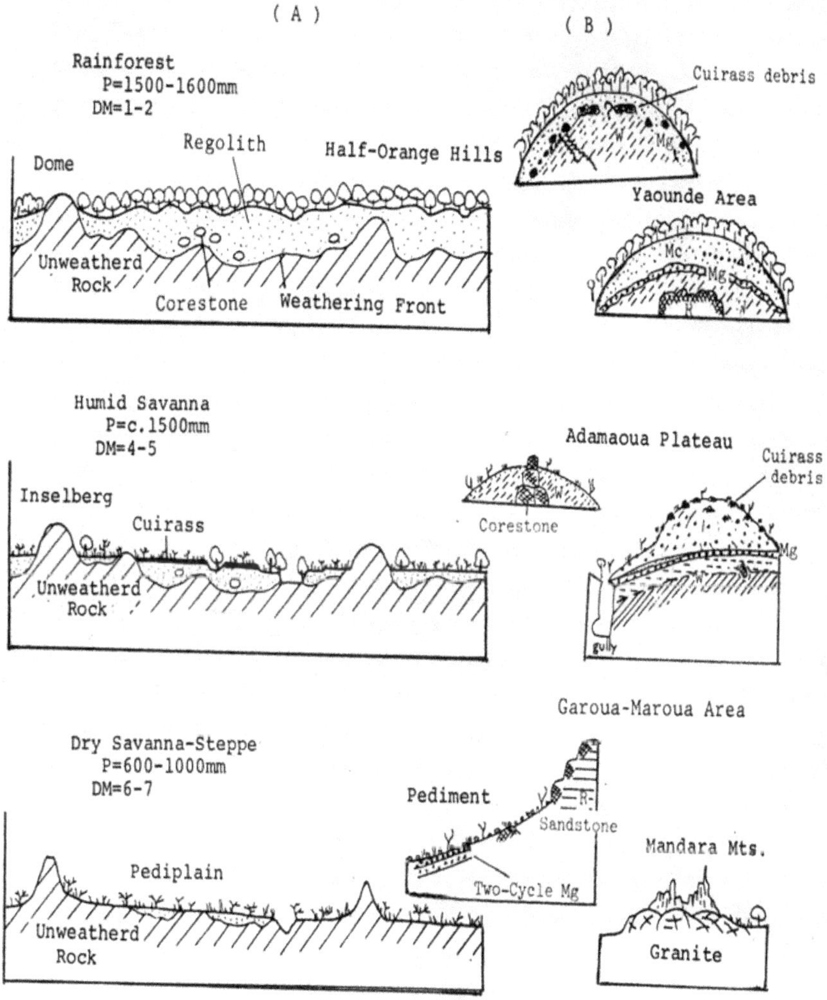

Figure 4b: *Regional variation in landscape in Cameroon according to climatic and ecological conditions. (A): modified from Demangeot (1976) which is based on Thomas (1974); (B): modified from Kadomura(1977) P: mean annual rainfall (1931-1960); DM: number of dry months with monthly rainfall of less than 50 mm. For symbols of regolith profile see Figure 4C. Note: cuirass is laterite.*

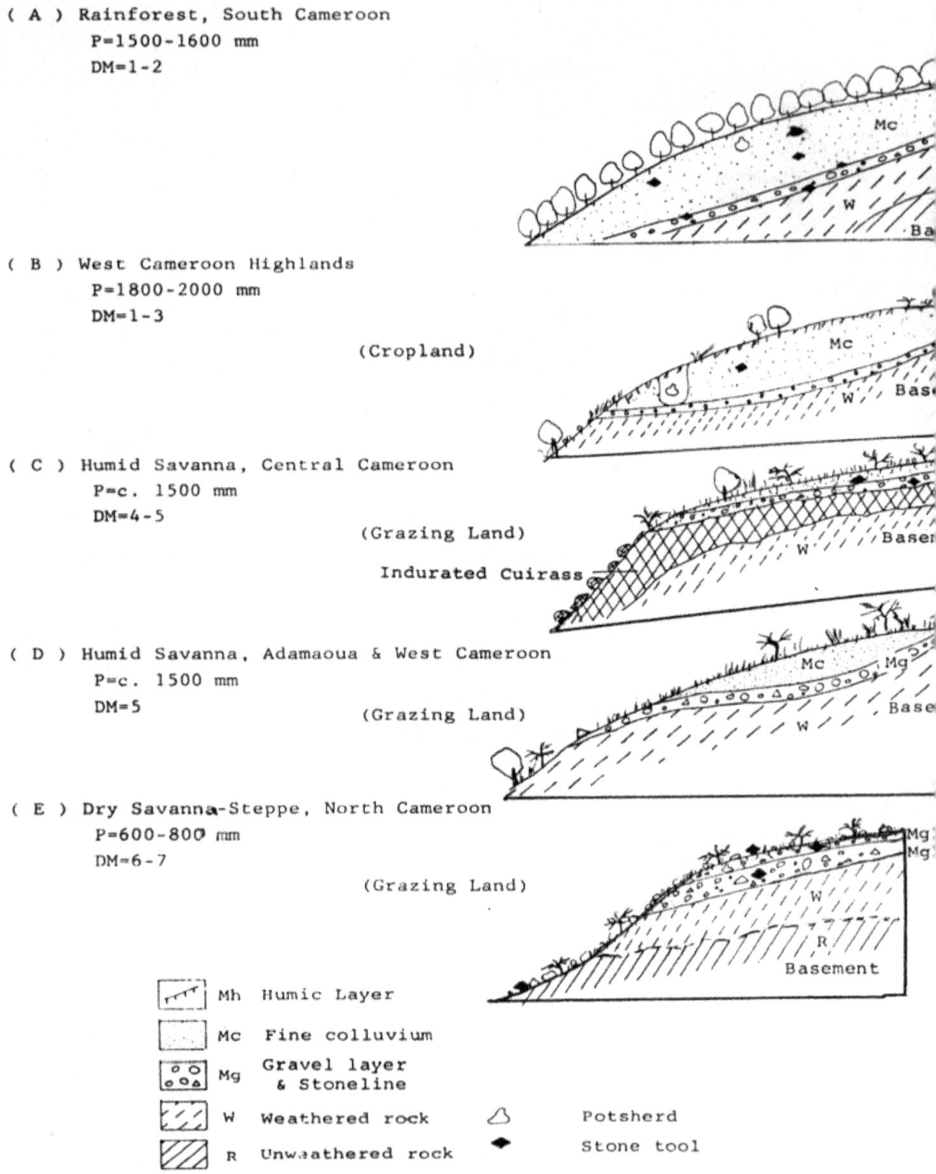

Figure 4c: *Regional variation in superficial deposits / regolith profile in Cameroon according to climatic and ecological conditions. Based on (A): Hori (1977a, 1977b, 1982); Kadomura (1977); (B): Tamura (1984, 1986); (C): Kadomura (1982a); (D): Tamura (1986); (E): Kadomura (1982b). Mgl: gravelly topsoil; Mgl1: second gravel layer. Note cuirass is laterite.*

The above forest degradation scenarios have provoked a savannization of the area once colonized by the forest. The emergence of grasslands and wooded grasslands attract pastoral enterprises. In deforested area the soils are characterized by an exposed layer of laterite on upper slopes and the topography by large gullies piping into forest galleries. These certainly reflect repeated cycles of clearing, burning and cultivation. Large fossilized termite mounds which exist in these areas are probably the remnants formed during the last glacial age. These are found in deforested areas. The impact of grazing and annual burning on the surrounding forest areas, that is, in forest-savannah contact zones has caused an accelerated degradation of the land. This is mainly due to soil erosion. These processes are schematized in Figure 5.

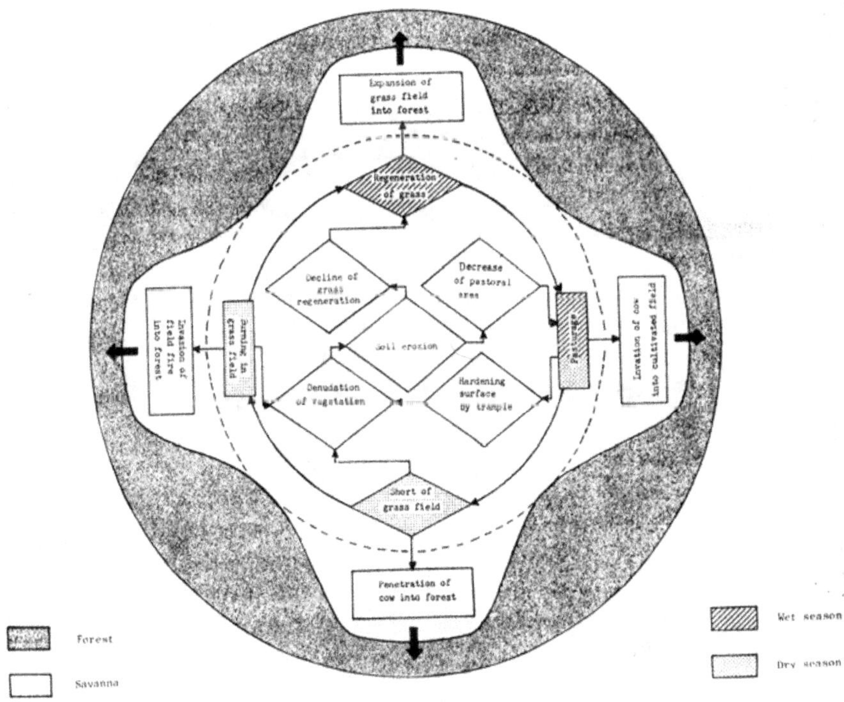

Figure 5: *Pastoral impact in forest-savannah contact zone.*

The earnings from export taxes remaining inside the country are reinvested only to a limited extent on rational economic measures benefiting forest conservation. The timber exploitation companies, on their part, make investments only in the short to medium term, on the principle of short-term profits. Few of them think along the lines of long-term forest utilization, orienting their investments accordingly. Timber exploitation is a brief pioneer phase leading to other forms of land use (cocoa and coffee plantations). This can be criticized because these land utilization types at current world market prices do not result in higher returns per hectare than would be possible under continued natural forest management.

Tropical timber utilization has clearly led to the accumulation of value for the benefit of state elites, the state and foreign exploitation firms, to the detriment of the original forest inhabitants, Cameroonians and notably the natural rainforest ecosystem and overall quality of the global environment.

Photo: *Vegetation dynamics in forest-savannah contact zones; grassland derived from sub-montane forest in Bafang*

Forest Management

Within the context of sustainable development of natural resources and the participation of the population in the management of forest resources Law No. 04/01 of January 20th, 1994 was passed to regulate forest exploitation. The strategies developed are:

- **Ensuring the protection of the forest heritage and preserving biodiversity.**
    - By creating national forest reserves and protected areas representing the national biodiversity.
    - By integrating the "environment and ecological equilibrium" concept in regional development policy.
    - By developing protection, improvement, and conservation measures for all forest resources.
- **Improving on the integration of forest resources in rural development:**
    - By promoting the preservation of forest resources and the domestication and raising of game in rural areas.
    - By promoting the development of private forests.
    - By promoting conservation of forest resources by local authorities.
    - By developing agro-forestry in farming systems.
- **Ensuring forest resources development:**

  Strategies aim at obtaining a better yield from forest resources. They include:
    - Increasing the supply of fuel wood and timber while maintaining the forest potential through a more rational use of available resources
    - Orientating demand towards less energy consuming methods or towards alternative energy sources; in the timber sub-sector, the strategies aim at:
    - Improving on the present procedures for obtaining forest exploitation licenses and the payment of exploitation fees;
    - Improving on the management of forest raw materials in order to optimize the productive potential;
    - Ensuring the regeneration of forest resources;
    - Promoting the use and the diversification of finished products.

- **Revitalizing the forest sector by setting up an efficient institutional framework:**

  The main strategies involved are:

  - The overall definition of the tasks of those involved in this sector.
  - Improving on the organization and coordination of the institutions involved in the use of forest resources.
  - Promoting training adapted to the objectives of the new forest exploitation policy.
  - Improving on the management of human resources.
  - Assisting in the development of the forest sector and financing the sector as a priority.
  - Periodically evaluating the implementation of the new forest exploitation policy.

From the above management strategies, the Cameroon Forestry Management Plan apart from omitting natural systems such as the mangrove ecosystems, riverine systems do not accommodate the observed and anticipated physical and biological effects of climate change on natural systems represented in the country. The combined effects of anthropogenic factors and climate change impacts require public awareness and political reactions. The savannization of forest-savannah contact zones in the country is the combined effect of climate change and landscape anthropogenesis.

Climate Change Vulnerability Analysis of Rainforests

The country has three climate regions (Figure 6). These are the humid-wet equatorial region in the south, a montane equatorial climate in the West, and a dry semi-arid climate in the North. Rainfall decreases from south to north. Figure 6 presents the actual rainfall and temperature regimes. There are high year-round temperatures and the weather is controlled by equatorial and tropical air masses. It has two major seasons, a dry season and a wet season. Most of the rain occurs between April and October. Rainfall is variable and unreliable (Ngakfumbe, 2001) and floods and droughts occur frequently.

The southern part of the country has an average temperature of 25^0C while the northern regions between 27^0C and 34^0C. The northern parts have seasonal rivers. These dry up during the dry season. The southern regions have abundant surface water resources. These drain into the Congo Basin; Niger Basin and Gulf of Guinea while the northern rivers but for the Benue River drain into the Chad Basin.

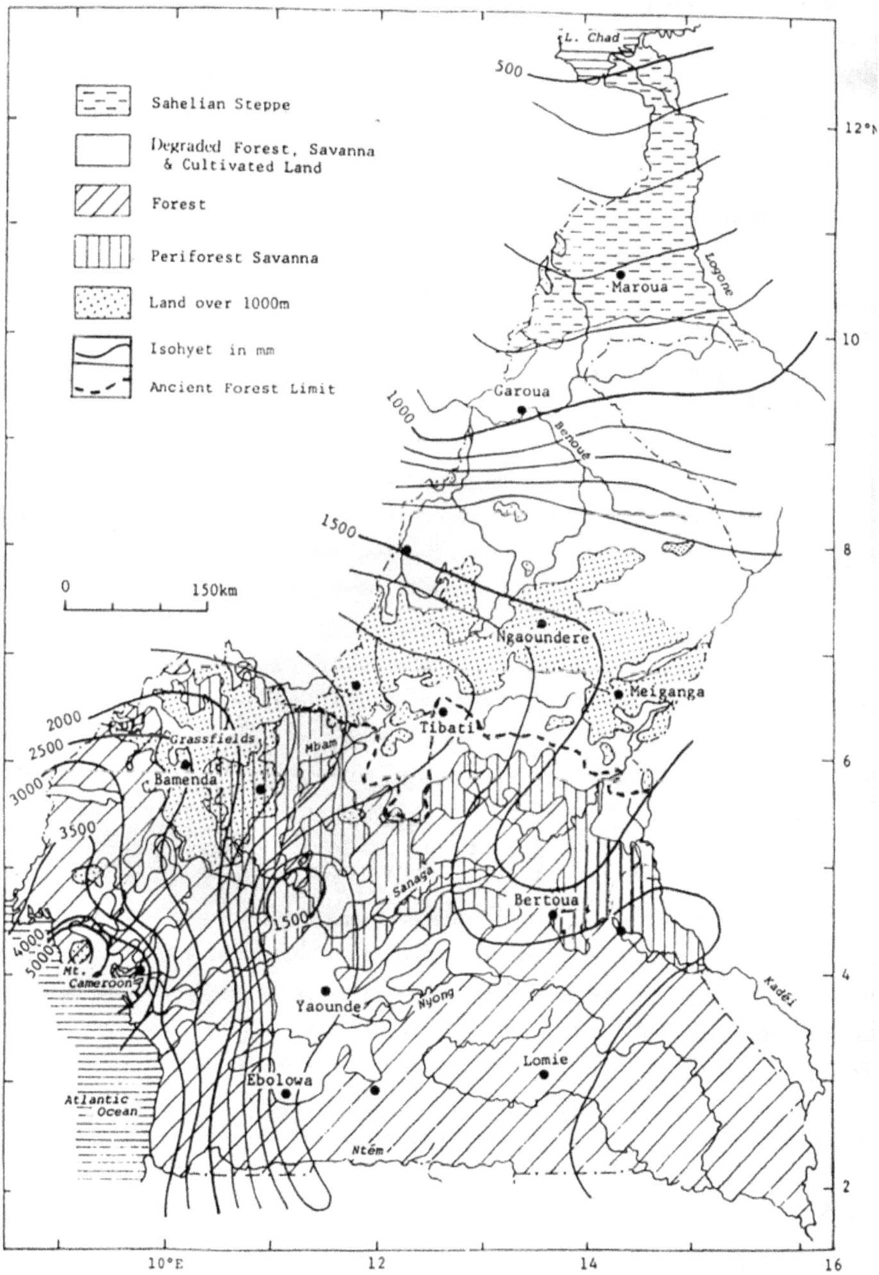

Figure 6: *Vegetation and climate in Cameroon*

The impact of climate change in tropical forest can be significant due to their relative sensitivity to climate variables. Corlett and LaFrankie (1998) established that phenological processes are highly correlated with climatic signals. Changes in plant phenology are said to be the most immediate indicators of climate change. Related to these are the co-evolutionary relationships that are highly specialized and could become disrupted if a species is affected by a change in phenological event or physiologically (Bazzaz, 1998).

The rainforest exhibits a high diversity of species per unit area. This has created narrow niches within the biome. These may be severely impacted by a reduction in biodiversity through elimination of species when migration to other suitable habitats is not an option, (Bazzaz, 1998). Shifting forest-savannah boundaries are certainly disruptions of this nature (Figure 6).

Due to existing levels of deforestation, the sensitivity of hydrological regimes will be exacerbated by climate change. In low-lying areas floods may increase in frequency and intensity (Bonell, 1998).

Fire is naturally rare or nonexistent in the rainforest. The interrelationships of fire, climate and forest ecology are complex. The frequency and intensity of bush fires are likely to increase due to climate change. Increased drought and the savannization of original forest lands could cause desiccation of forest. This could spur more fires. The tropical rainforest is not resistant to fires. Such events can therefore cause a large-scale loss of biodiversity. The forest will experience a change in species composition that will favour xenomorphic, pyrophytic and generally species-poor plant communities that create a positive feedback as the new ecosystem will be increasingly liable to burn. These processes are presented in figure 5. They are exacerbated by human factors.

WWF (2003) citing Hilbert (in press) estimates that a rise in temperature of 1^0C would increase the productivity of rainforests as a whole as long as all other factors are held constant. However, changes in rainfall patterns combined with warming can produce sizeable shifts in the distribution of forests types in the country. If we assume the absence of fire, no pests and no effects of other factors, increased rainfall will enhance the productivity of the forest. On the other hand, decreased rainfall could shift existing rainforest boundaries in favour of savannah woodlands (Figure 6) and probably accelerated deforestation of the northern regions of the country. Shifts in distribution are expected to take place at the edges of forest types and in ecotonal areas between rainforest and more open forest (Hilbert, 2001).

Dudley (1998) provides an overview of major categories of changes expected in forest ecosystems:

- **Disturbance:** climate change will increase the degree of disturbance, through extreme whether events such as storms and as a result of smaller but ultimately pervasive changes in seasonality, rainfall and temperature. Climate change will thus add to those other forms of human disturbance, which are currently fragmenting and altering forest ecosystems.
- **Simplification:** The net effect of problems with tree reproduction and species' migration rates in areas experiencing climate change will tend to cause problems for slower growing species and instead favour fast growing species, short-lived weed and invasive species. The result will be an acceleration of a trend that is already occurring as a result of other forms of human interference, namely, the replacement of species-rich forest by species-poor forests.
- **Movement:** Movement is likely both geographically and attitudinally, as growing conditions alter. The ability of trees to migrate fast enough to keep pace with climate change is still largely unknown and will depend upon many other factors. The extent to which ecological conditions change will depend on a complex mixture of factors; for example, warmer conditions could encourage trees to move up-slope while accompanying droughts might have the reverse effect.
- **Age reduction:** Disturbance, increased forest fires, changes in pest patterns and the transition of whole communities will encourage an existing trend towards the replacement of old-growth forests with younger stands. This has particularly important implications for biodiversity, as many of today's threatened species are those confined to older habitats. Examples are the paleo-endemics of the rainforest at the foot slopes of Mount Cameroon.
- **Extinction or Extirpation:** Some of the most vulnerable forest habitats, including relic species at the edge of their ecological niche and some particularly threatened systems, such as the mangroves in low-lying coastal regions, could disappear all together. Species could also disappear from some forests that appear to be surviving the changes relatively well.

Adaptation Options to Promote Resiliency

Long-term adaptation of species will be enabled where natural adaptation processes such as migration, selection and change in structure are allowed to take place due to sufficient connectivity and habitat size within the landscape. Traditional forest conservation methods must emphasize an increase in the temporal and spatial scales, protection of key forest communities, managing forests specifically for increased disturbances, and flexibilities given uncertainties and surprises surrounding what climate change will bring (MPWG, 1998). Sustainable forest management criteria provide a framework into which adaptation strategies can be incorporated (NRC, 2002):

- **Reduce forest threats**: strategies should seek to promote overall ecosystem health. Anthropogenic threats must be minimized. These should be targeted for priority action to ensure that forest structure, composition and function are increased in order to increase resilience.
- **Avoid fragmentation and provide connectivity**: Forest fragmentation present "Edge effects" which threaten the microclimate and stability of a forest as the ratio of edge to interior habitat increases. This reduces the ability of the forest to withstand delibitating impacts. It also contributes to loss of biodiversity as exotic, weedy species with high dispersal capacities are favoured and many native species are inhibited by isolation (Noss, 2000).
- **Maximize size of management units**; **Biogeographic scale:** Fossil records in the West Cameroon Highlands provide evidence that species have adapted to changing climates by shifting their boundaries attitudinally and southwards. Decision-making to determine the boundaries of protected areas should therefore bear this in mind. Attention should be given to the need for north-south migration and altitudinal migration opportunities. Land use plans must therefore consider such biogeographic scales.
- **Provide buffer zones and flexibility of land uses**: in Cameroon the boundaries of protected areas are fixed. Such fixed boundaries are not well suited to a dynamic environment. Protected areas are also very small in size. Hardly of any biogeographic scale. These offer little room for the creation of buffer zones. With changing climate, buffer zones might

provide suitable conditions to accommodate pressure from human populations.

- **Represent forest types across environmental gradients**: Set aside various forest types for scientific study in order to monitor disturbed areas, and as a means of conserving species in protected areas. This should enable some resistant and resilient types to persist (Noss, 2000).
- **Protect functional groups and keystone species**: the aim should be to maintain the natural diversity of species and functional groups in a forest. Functional groups include various kinds of producers, pollinators, seed dispersers, predators, parasites, decomposers, and so on. Species diversity and composition enhance stability. Keystone species should also be identified and protected and their natural patterns of distribution and abundance maintained. This should increase resistance and resilience to climate change.
- **Maintain natural fire regimes**: The frequency and intensity of fires are known to correlate with changing climatic conditions. Promote the integration in management strategy the development of "fire-smart landscape", that reduce the spread and intensity of wildfire (Hirsch *et al.*, 2001).
- **Other strategies to increase resistance and resilience include**:
  - The use of silvicultural techniques to promote forest productivity;
  - Prevent the conversion of rainforest to agro-industrial plantations and practise low-intensity forestry;
  - Maintain genetic diversity and promote ecosystem health through restoration of degraded forest areas
  - Assist migration with species introduction in new areas of the rainforest zone of the country;
  - Protect most highly threatened species ex-situ in botanical gardens as it is currently being practiced in the Limbe Botanical Garden for both plants and animals.

Conclusion

The above adaptation strategies are key measures for the short and the medium term. These measures will increase resistance and resilience to climate change only if the negative impacts of anthropogenic factors are arrested. While forest resource managers confront the blend of adaptation strategies necessary to mitigate climate change impacts, there is the urgent need to control the human-induced changes in temperature and rainfall. Therefore, the only long-term option to ensuring healthy rainforest ecosystems is to reduce the emissions of greenhouse gases.

References

Anon (1998) The rape of the rainforest. *Awake Magazine*. Watch Tower Society. Printed in Germany

Ayonghe, S. (2001) A quantitative evaluation of global warming and precipitation in Cameroon from 1930 to 1995 and projections to 2060: effects on the environment and water sources. In: C.M. Lambi and E.B.Eze (eds) *Readings in Geography;* Unique Printers, Bamenda. p. 142-155.

Bazzaz, F. (1998) Tropical forest in a future climate: changes in biological diversity and impact on the global carbon cycle. *Climate change, vol. 39*, p. 317-336.

Besong, J. and Ngwasiri N. (1995) The 1994 forestry law and rational natural resource management in Cameroon. *PVO-NGO/NRMS Cameroon Publication*, Yaounde. P. 2-5.

Bonell, M. 1998) Possible impacts of climate variability and change on tropical forest hydrology. *Climate change, vol 39,* p. 215-272

CAMPER/USAID (1993) USAID-Cameroon Programme for Environmental Reform. CAMPER.

Carter, R. and Alkali, A. (1996) Shallow groundwater in the northeastern arid zone of Nigeria. *Quarterly Journal of Engineering Geology, vol. 29,* p. 341-355.

Corlett, R. and LaFrankie, J. (1998) Potential impacts of climate change on tropical Asian forests through an influence on phenology. *Climate change, vol 39*, p.439-453.

Dudley, N. (1998) Potential impacts of climate change on forests. A Report for WWF International

Gartland, S. (1994) Cameroon, In. Sayer (ed) *Conservation Atlas of Tropical Forest, Africa.* J., Harcourt C., and Collins M. London.

Guiffo, T. (1994) Nouvelle Législation Forestière au Cameroun. Foundation Friederich –Ebert, Yaounde.

Haruki, M. (1984) Silviecological studies in the savannah and forest zones of Cameroon. In: H. Kadomura (ed). *Natural and man-induced environmental changes in tropical Africa: case studies of Cameroon and Kenya.* Hokkaido University, Sappro. P. 75-91.

Hilbert, D.; Ostendorf, B; Hopkins, M. (2001) Sensitivity of tropical forests to climate change in the humid tropics of North Queensland. *Australian Ecology, vol. 26*, p. 590-603.

IUNC (1986) Review of the protected areas system in the Afrotropical realm. IUNC, Gland. 259p.

Jurgen, P. (1990) Contribution to the development of timber exploitation and wood processing industries in countries of the humid tropics: case of Ivory Coast. *Natural Resources and Development.* Inst. of Scientific Cooperation. Tiibingen.

Lanly, J-P (1982) Tropical forest resources. *FAO Forestry Paper No. 30* FAO Rome.

Miller, G.T. (1994) Sustaining the earth: an integrated approach. Wadsworth Publishing Company, Belmont California

MINEF (1996) Environment and sustainable development for Cameroon. MINEF, Yaounde

MINEF (1992) National Environmental Management Plan, Vol. 1, Main Report, Yaounde.

MPWG (1998) The Montreal process, assessed August 2003 on-line at http://www.mpci.org/criteriaehtml. The Montreal Process Working Group.

Ngakfumbe, S. (2001) Rainfall probability and reliability: the case of Cameroon. In: E.B. Eze and C.M. Lambi (eds). *Readings in Geography.* Unique Printers, Bamenda, p. 151-175.

Noss, R. (2000) Managing forests for resistance and resilience to climate change. A report to World Wildlife Fund, USA

NRC (Natural Resource Canada) (2002) Climate change impacts and adaptation: a Canadian perspective. Prepared by Climate Change Impacts and Adaptation Directorate. Assessed August 2003 on-line at http://adaptation.nrean.gc.ca/app/filerepository/FDE8A92C21A248CDB135F6373D7ED5C5.pdf

WWF. (2003) User's manual for building resistance and resilience to climate change in Natural systems. WWF Climate Change Programme, Washington D.C. p. 175-214.

Chapter Three

Climate Change and Mangrove Ecosystem Services

Summary

Mangrove swamps are a pantropical ecosystem common along the West African coast where they account for 64,623 square kilometres, about 3.1 percent of the region. It is rapidly being converted to other land use systems with complete disregard to its actual and potential value to coastal living resources due to neglect and their being considered as waste land by governments. These are also threatened by droughts and a rise in sea level resulting from global warming. The chapter examines the distribution, structure, composition, direct and indirect use values and its vulnerability to the mix of natural events triggered by climate change and anthropogenic factors. The chapter posits that despite the multifunctional resources and ecosystem services they possess, mangrove forests have been considerably misused and mismanaged. As a result of anthropogenic degradation these resources and services risk being lost completely. Climate change will trigger a modification of habitat factors such as soil conditions, temperature, rainfall, topography, salinity, depth of water, tidal range and turbidity. It therefore concludes that the mangrove ecosystem is vulnerable to climate change, agricultural and industrial pollution and unsustainable management. The consequence is the loss of ecosystem multifunctional resources and services due to the delicate balance of its ecosystem components. Finally, the chapter concludes that the mangrove ecosystem can play an essential role in mitigating climate change impacts through carbon sequestration, coastal zone protection from storm surges and floods, stabilization of the coast through accretion, and in buffering the influence of climate change as well as supporting local communities in adapting to it. Therefore, it devises strategies for mitigating the mix of anthropic and climate change impacts.

Key words: Anthropic impacts, climate change impacts, mangrove forests, ecosystem services, multifunctional resources, habitat factors, mitigation.

Introduction

There is a great potential for global climates to change as a result of increasing green house gases and their effects, that is, the effects of global warming. These are being manifested in the form of droughts, floods, rise in temperatures and a thermal expansion of the oceans by melting middle and high latitude glaciers and ice sheets leading to a rise in sea level. These are likely to alter the vulnerability and frequency of events related to droughts, storms and floods with serious adverse impacts on biodiversity and ecosystem services, that is, the combination of life forms and their interactions with one another and with the physical environment that has made earth habitable for man. Ecosystems provide the basic necessities for life, offer protection from natural disasters, and are the foundation for human culture. However, the mangrove ecosystem services are being lost as a result of a mix of anthropogenic and natural factors related to weather changes. Climate change has a major impact on biodiversity and therefore also on ecosystem services such as food and water provision. At the same time, biodiversity in the form of forests, functional soils or genetic resources can play an important role in mitigating climate change, buffering the influence of climate change as well as supporting communities in adapting to it (Vohland, 2008).

Mangrove swamps are trees or bushes growing in the intertidal zone of sheltered tropical and subtropical shores. This community colonizes about 25 percent of the world's tropical shorelines (Sasekumar, 1978). These swamp forests are almost always erroneously regarded as wasteland believed to be infested with snakes and other vicious creatures. Some regard mangroves as "freaks" and many government authorities regard them as wasteland. Consequently, among the many environmental degradation problems facing the world today is man's interference of the coastal zone. This impact has resulted in the pollution and contamination of the coastal waters, and the critical modification of the coastal ecosystems (Ndenecho, 2008). With the advent of industrialization, urbanization and commercial agriculture in Africa, mangrove coastlines have come under increasing pressure for development. The purpose of this review is to examine the distribution, structure, composition, multifunctional resources and ecosystem services and potential climate change and anthropogenic impacts on the mangrove forest. The current quantity of green house gases in the atmosphere is due to a mix of natural and man-induced events (Pearce and Warford, 1995). While the impacts of the man-induced events can be reduced through sustainable biodiversity management actions, there is need to mitigate climate change and assist communities to adapt to it. An estimated 100.000 km^2 of mangrove swamp forest are distributed across tropical regions of the world. In the Philippines more than 330.000hectares have been destroyed by agriculture and aquaculture. In Ecuador 60.000 to 70.000hectares have disappeared following the development of farming. In West Africa many of the mangrove forests have been cleared for agriculture, urban development and for timber products (Kuete, 1998).

The Ecology of Mangrove Forests

The ecology of the mangrove forest has been described by Ndenecho (2007), Sasekumar (1978), Spanniks and Van Beukering (1979), Odum (1970) and Millington *et al.* (1994). The main ecological aspects of the West African communities are:

- **Geographical Distribution:** Mangrove swamps are common along the West African coast, but are particularly important in these areas (Figures 1 and 2):

  - From the Casamance River in Senegal to the Rio Nunez in Guinea, a coastal strip that includes the extensive mangrove of the Bijago's Islands and the Cacine, Cacheu, Corubal, and Gêba rivers in Guinea-Bissau and the Komponi river in Guinea.
  - Around Conakry in Guinea.
  - In Sierra Leone, especially in the estuaries of the Scarcies, Rokel, and Jong rivers, and around Sherbro Island.
  - As a series of small isolated occurrences in Liberia and Côte d'Ivoire
  - In the Volta Delta in Ghana.
  - Along the Nigerian coast, especially in the Niger and Cross deltas.
  - Along Wouri, Sanaga and Ndian coast of Cameroon.

These areas account for 64,623 square kilometres, about 3.1 percent of the region. The largest areas are in Nigeria (35,200 square kilometres), Liberia (5,533 square kilometres), Guinea (5,480 square kilometres),

Cameroon (750 square kilometres) and Ghana (5,111 square kilometres). The hinterland is colonized by the rain forest which in many countries is rapidly being degraded by human activities (Figure 1 and 2).

- **Structure and Composition:** Mangrove swamps occur only in tidal and brackish water along the coast in places where adequate shelter from storm waves usually exists. They commonly occur fringing wide estuaries, in shallow creeks, and behind islands. Mangrove is swamp forest in which a dense network of tree trunks, stilt roots, and pneumatophores usually form an impenetrable thicket. The tree roots are directly in the mud of the tidal flats. The canopy cover is high and mostly evergreen, generally with no ground cover.

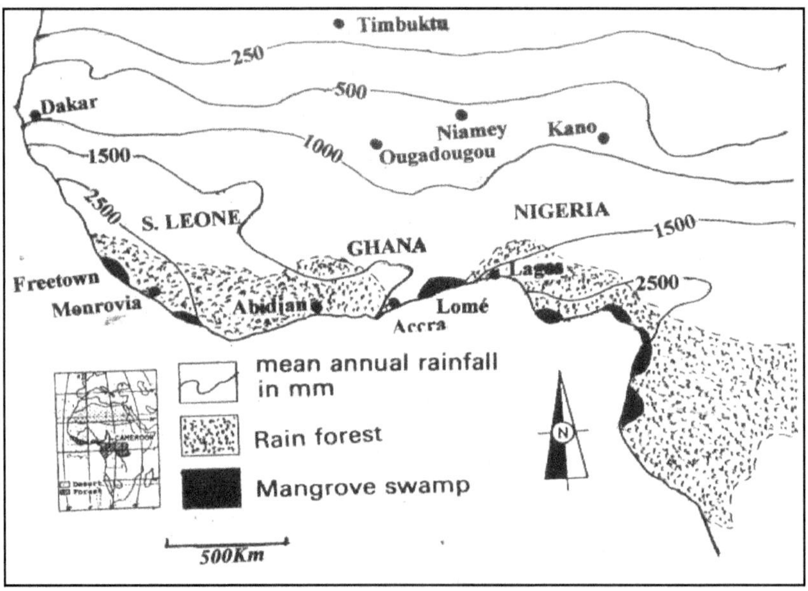

Figure 1: *Location of remnants of Mangrove forest and broad leaved forest in West Africa. Mangrove forest Area =64,623 km² or 6,462,300 hectares capable of storing 746,623,800 tonnes of carbon. 1 hectare of closed undisturbed broad-leaved tropical forest stores 117 tonnes of CO_2 (Pepper, 1992). CO_2 storage capacity = 75,608,900 tonnes.*

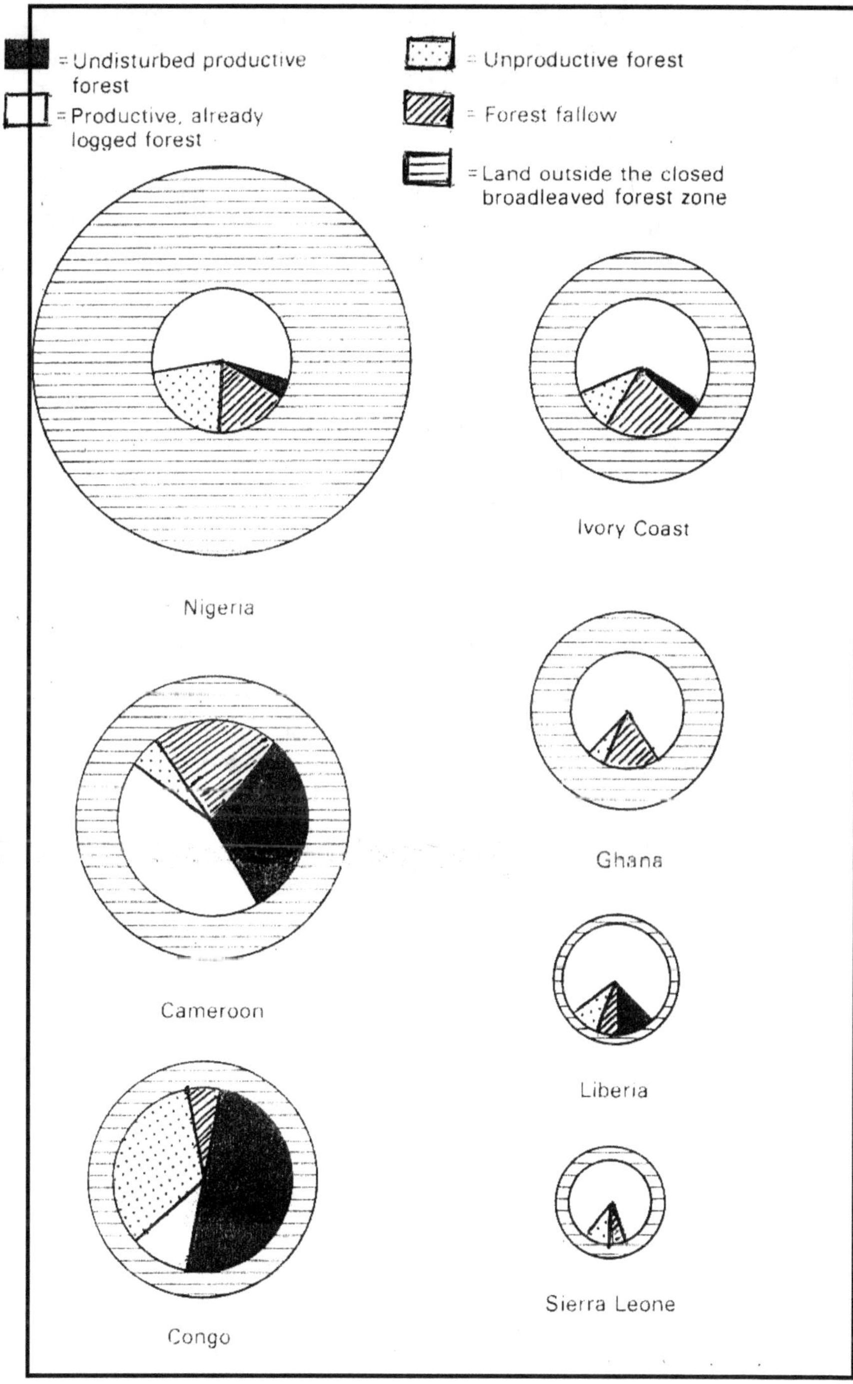

Figure 2: *Differences between countries in West and Central Africa in the area of closed, broadleaved forest, and the extent of its conversion (Longman*

and Jenik, 1987). A broadleaved closed tropical forest is capable of storing 117 tonnes of carbon per hectare. (Pepper, 1992).

The mangrove forests of West Africa are dominated by three genera: *Avicennia, Laguncularia, and Rhizophora.* Spatial relations among the genera, however, are not always clear. Mangrove trees of the genus *Avicennia* generally occur inland of *Rhizophora.* However, along the Gambia River, for instance, no obvious zonation occurs (Giglioli and Thornton, 1965), and the Niger Delta is dominated by a mixture of *Rhizophora harrisonii,* R. *mangle, R.racemosa, and Avicennia gerninans.*

In Sierra Leone, mangrove swamp zonation is thought to be controlled by soil conditions, topography, and water salinity (Cole, 1968). In Sierra Leone, *Rhizophora racemosa* is the pioneer species on silty, fibrous soil, while on sandy soils the pioneer species are *Avicennia Africana,* R. *mangle*, and R. *harrisonii.* Trees at the mouths of rivers attain maximum heights of 7.5 to 10meters, while upriver, at the limit of tidal incursion, they grow to 20 meters.

Inland from the area of pioneer trees, the open woodland changes to dense, thicketlike swamp forest often dominated by *Conocarpus erectus* and *Laguncularia racemosa.* In these areas, grasses and sedges characteristic of freshwater swamps invade to form an undergrowth, whereas other areas display a mixture of mangrove and freshwater swamp forest tree and shrub species.

- **Mangrove Productivity and biomass:** Mangrove productivity and biomass are very closely correlated with mean annual rainfall. The greatest productivity exists in humid tropical forest areas, although they show only moderate seasonality. This relation can be illustrated using tree height as an indicator of the growing stock. In the Niger Delta, where the mean annual rainfall is more than 4,000 millimetres, *Rhizophora mangle* stands attain 45 meters in height. (Millington *et al*; 1994). But at the northern and southern limits of mangrove on the west coast of Africa, tree heights are much lower. At the northern limit on Ile Tidra, Mauritania, $19^{0}50$'N, where mean annual rainfall is 100 millimetres, *Avicennia* spp. trees are only a few meters high (Chapman, 1977). At the southern limit in Benguela, Angola, $12^{0}30$'S, where mean annual rainfall is approximately 150 millimetres, mangrove again is quite stunted (Barbosa, 1970).

Manifestations of climate change are increasing temperatures and variability in precipitation resulting in floods and droughts. Rising sea levels will therefore negatively affect coastal ecosystems. Mangroves are highly specialized trees of the coastal region which are adapted to high levels of brackish water. The plants harbour a variety of species including commercial ones, and provide local communities with mechanical protection against inundations and tsunamis.

Table 1: *Mangrove area, growing stock, sustainable yield and CO_2 storage potential in West Africa (Figure 1)*

Location of Mangrove forest	Area		Growing stock		Sustainable stock		Potential to store CO_2 *
	Km²	Percent (%)	Tonnes (000)	Percent (%)	Tonnes/year (000)	Percent (%)	Tonnes/ha (00)
Ghana	5,111	2.16	126,139.4	21.80	15,072.3	67.71	597,987
Benin	738	0.65	18,213.8	6.68	2,176.3	33.39	86,346
Guinea-Bissau	2,793	9.28	68,931.2	58.85	8,236.5	91.78	326,781
Guinea	5,480	2.29	135,246.4	20.76	16,160.3	66.53	641,160
Ivory Coast	3,741	1.15	92,327.8	13.39	11,023.2	58.44	437,697
Liberia	5,533	5.88	136,534.4	47.10	16,316.8	89.37	647,361
Nigeria	35,200	4.00	868,736.0	36.03	103,804.8	74.51	4,118,400
Sierra Leone	4,592	6.53	113,330.5	49.98	13,541.8	90.68	537,264
Togo	685	1.27	16,905.8	13.63	2,020.1	53.76	80,145
Cameroon	750	/	/	/	/	/	87,750
Total potential capacity to store carbon dioxide							756,089

Source: Milington et al. (1994). * Author's calculations after Pepper (1992).

Table 1 summarizes the area covered by mangrove forest, the growing stock and sustainable yield for the West African Coast nations of Benin, Ghana, Guinea – Bissau, Guinea, Ivory Coast, Liberia, Sierra Leone and Togo. The growing stock is air-dried, above the ground biomass, expressed in tonnes per hectare. The sustainable yield is the mean annual increment of air-dried, above the ground woody growth, expressed in tonnes per hectare per year. Mangrove land cover in West Africa is very limited and the growing stock is generally low. Despite the economic value and ecosystem services it provides the ecosystem has been misused and mismanaged. With a total land cover of 64,623km^2 it has a potential to store 75,608,900 tonnes of CO_2 (Table 1 and Figure 1). The destruction of this ecosystem can therefore result in the discharge of enormous quantities of CO_2 into the atmosphere. Being one of the greenhouse gases, this will accelerate global warming.

Many of the mangrove swamps in the region have been cleared for swamp rice cultivation. This clearance has had an important impact on woody biomass stock, as is well illustrated by Sierra Leone. Although Temne farmers cultivated swamp rice in northern Sierra Leone estuaries in the nineteenth century, the main impetus to clear mangrove swamp vegetation for cultivation came with twentieth century colonial penetration of the interior (Millington, 1987). Clearance of mangrove swamps leads to a buildup of sulfuric acid to toxic levels in the thionic fluvisols of the swamps. This acidity initially restricted rice cultivation, but research by the West African Rice Development Association (WARDA) in Sierra Leone on soil management techniques and the introduction of specific rice varieties has led to the widespread use of mangrove swamps for rice farming.

Due to the concentration of population along the West African coast, pressure on mangrove swamps can locally be very great. Pressure arises from land clearance for rice cultivation, from fuelwood exploitation, from the preferential use of mangrove wood for smoking fish and baking bread, and from its use in building construction. Areas of mangrove adjacent to towns suffer severe exploitation, and in these areas fuelwood supply problems undoubtedly exist, accentuated by the value placed on mangrove wood because of its high calorific value.

Mangrove ecosystems are multi-functional resources. They provide a wide range of goods and services. Use values can either be direct or indirect. Direct use value is derived from direct use or interaction with mangrove resources and services. Indirect use value is the support it provides to off-site fisheries through their nursery function, carbon sequestration, protection of coastal zones and stabilization of areas of high accretion. Today, the provision of ecosystem services is further endangered due to rising sea levels. Flood and droughts combined with an overall loss of biodiversity not only due to climate change, but also due to overexploitation, land use change and chemical pollution threaten these ecosystem services.

Mangrove Ecosystem Services

The Millennium Ecosystem Assessment was the first global assessment of changing "ecosystem services" as consequence of global environmental degradation (Vohland, 2008). Ecosystem services describe functions of ecosystems that are available for man and human development. Climate change and anthropogenic pressure will trigger a modification of the mangrove ecosystem and loss of ecosystem services. Biological diversity is crucial to sustaining human livelihoods. Poor people are particularly dependent on continued biodiversity. The great wealth of plant and animal species and of the mangrove land is the basis of their livelihood, their food supply and – through medicinal plants their health. The mangrove forests provide the following ecosystem functions:

- **Timber products:** The wood obtained from mangrove is good fuel as charcoal or firewood. The timber is extensively used in house construction, particularly in coastal towns and villages. Fuelwood is used for cooking and smoking fish. Timber products are marketed (Youmbi *et al*, 1997; Fondo, 2001).
- **Livestock feed:** Mangrove leaves are used as livestock feed.
- **Medicine:** Mangrove leaves are used for various medicinal purposes.
- **Tannin:** Mangrove bark is a commercial source of tannin used in the tanning of leather and preservation of fish nets.
- **Mangrove land:** Mangrove land has been reclaimed for agriculture, especially in the deltas and creeks. The pioneer crop in these areas is rice which requires temporary waterlogged conditions. Rice cultivation is successful, provided salinity can be suppressed during the growing season by adequate drainage. Large areas have been cleared to produce rice and shrimps. Now their provision of ecosystem services is further endangered due to rising sea levels. Rainfall variability resulting in extreme drought and flood events will alter the natural ecological balance of acidity and salinity in the swamps. As plants and animals rely on specific environmental conditions, changes in temperature and humidity regimes affect the performance and competitiveness and relationships of species.
- **Mangrove dependent fisheries:** Mangrove ecosystems provide food and shelter for an infinite variety of coastal fishes and shellfish. The pathways by which energy derived from the mangrove ecosystem supports coastal and estuarine fisheries have been described by Heald and Odum (1969) and Odum (1970). This

detritus food web based mangrove plant material is presented in figure 3.

- **Provision of shelter for fish:** The mangrove ecosystem including shallow and calm rivers and creeks provide shelter for the juvenile stages of many commercially valuable prawns. A significant amount of the diet of juveniles and sub-adults that frequent mangrove waters consists of mangrove plant detritus. The mangrove is essential for the survival of the prawn fishery (Pathansali, 1976). The removal of mangrove from a locality can therefore bring about the depletion of offshore marine fisheries through loss of food and shelter for the juveniles of prawn and fish.
- **Protection and shelter:** Mangrove can afford protection to coasts from tidal waves. It has been suggested that the devastating effects of the 1970 tidal wave which struck Bangladesh would have been less severe if the coastal belt of mangrove forests had not been converted into paddy fields (Foresberg, 1971; Chapman, 1971).
- **Stabilization of areas of high accretion:** Mangroves can stabilize areas of high accretion and growth rates of mangrove shores may be substantial in regions of high sedimentation. Rates of accretion in sub-Saharan Africa may be greatly accelerated due to the anthropic degradation of inland areas. Substantial amounts of sediments derived from deforested areas are carried out to the sea by the many rivers. These sediments subsequently settle on the shore through tidal and wave action (Dixon, 1959). Figure 4 shows the edaphic effect of Mangrove forests on vegetation.
- **Carbon sequestration:** Pepper (1992), estimates that in Africa, 117tons of carbon is stored in a hectare of closed broadleaf – tropical forest which is undisturbed. Fondo (2001) estimates that 30hectares of mangrove in the Douala estuary in Cameroon are converted to other uses every year. The loss of this biomass therefore will increase the carbon concentration of the atmosphere and thereby global warming. Carbon trapped in the ecosystem does not contribute to expected future green house effects. According to Pepper (1992) in Africa, 117 tonnes of carbon is stored in a hectare of closed broadleaved forest. This is the case of above ground biomass in undisturbed forest. The West African Mangrove forests

can therefore store about 75,608,900 tonnes of carbon dioxide (Table 1 and Figure 1).

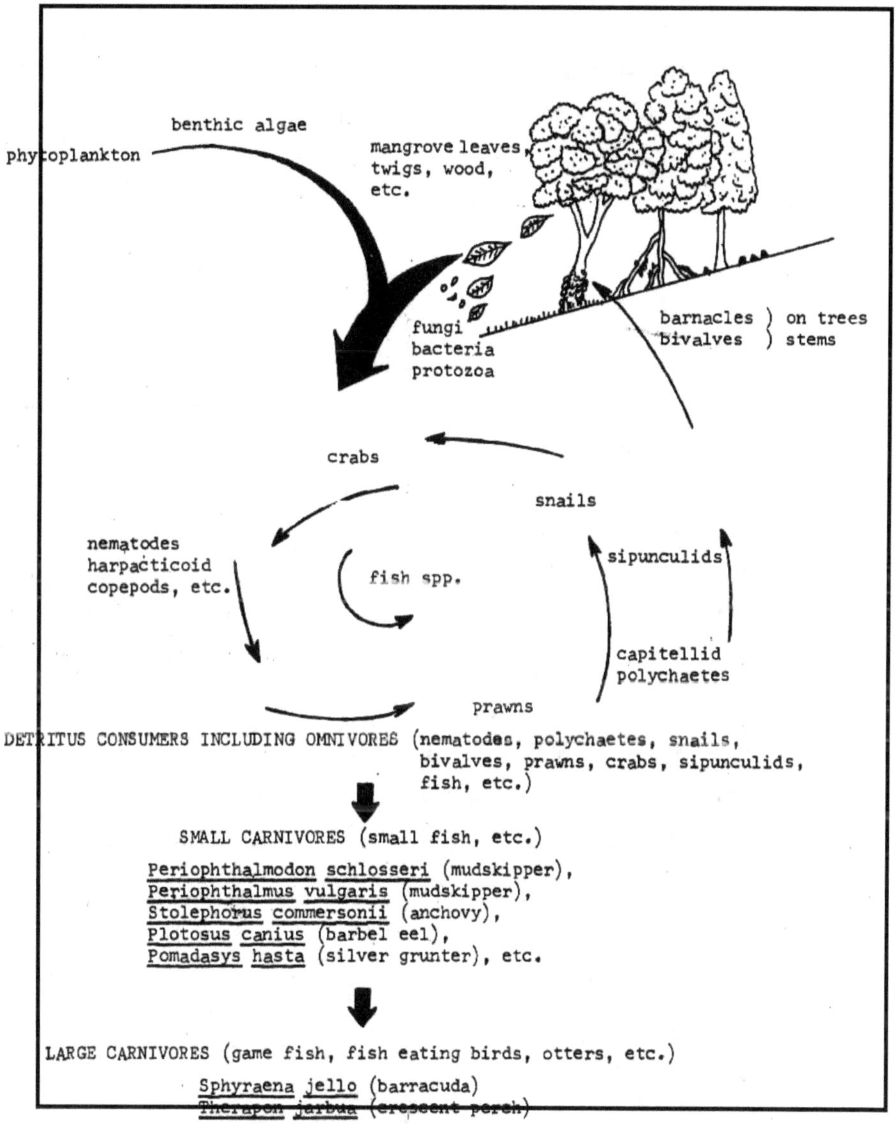

Figure 3: *The detritus food web based on mangrove plant material in the mangrove shore and adjacent waters of the west coast of peninsular Malaysia (modified after Odum, 1970)*

- **Non-use values of Mangrove forests:** Mangrove forests also have non-use values. These may arise from the satisfaction an individual derives from knowing that mangroves continue to exist, but is not necessarily planning to use them (Fondo, 2001). Another possible motive of non-use value is the desire to preserve mangrove ecosystems for future generations. In view of rural and urban poverty and the livelihoods mangrove ecosystems support, it should be very unlikely than non-use values for the mangrove forests in West Africa will be very important. People will not be willing to sacrifice for its preservation if its degradation does not present a serious threat to their livelihoods.

- **Option value of Mangrove forest:** Fondo (2001) defines the option value as an individuals willingness to pay to preserve the option of using a good in future. It is generally believed that as the level of scientific knowledge increases, more uses will be found for mangrove forest resources.

The above ecosystem services (indirect use value of mangrove ecosystems) and products (direct use value of mangrove ecosystems) are the benefits to be derived from the ecosystem. These may be lost in the event of climate change which could trigger a modification of the ecosystem. These are multifunctional resources. Some of the multiple uses may conflict with each other. Mangroves are rich ecosystems, capable of providing a range of goods and services for human populations.

Figure 4: *Photograph showing the edaphic effect on vegetation: Foreground = Mangroves are growing in the mud of a river mouth, Background = Coconut palms are growing in the sandy soil along the seashore.*

Habitat Factors in the Mangrove Environment

The mangrove ecosystem and its delicate web of life may be affected by the following:

Depth of water: Only small organisms can live in shallow water. Light penetrates to the bottom of shallow water so that plants can survive in it even at the bottom is an important factor. Atmospheric oxygen dissolves in water at the surface. In shallow water most of the water is well supplied with oxygen. During droughts shallow water supporting mangrove ecosystems may be subject to evaporation and drying out. In the case of rise in sea levels the deep water may be dark at the bottom and the survival of plants will be difficult.

Turbidity: Turbid or muddy water resulting from floods in case of sea level rise obstructs light penetration. Green plants in the mangrove environment may not survive below a certain depth in turbid water. Without plants, there will be no food for herbivorous water animals. Also oxygen will not be released into the water by photosynthesis, as would be the case if plants were present. So the oxygen available to plants will be reduced.

Tides: Tides alter the level of water along the seashore. The inter-tidal zone is alternately flooded and exposed twice a day. Climate change in the form of increased precipitation variability and droughts will displace the inter-tidal zone. Mangrove trees, seaweeds and animals of the inter-tidal zone which usually have adaptations for attachment and for withstanding exposure to dry air will be adversely affected.

Salinity: The body of fluids in organisms of the mangrove environment is usually in osmotic balance with salt water in which they live. Plants and animals of estuarine waters have to be able to withstand wide and sudden variation in salinity. As plants and animals rely on specific environmental conditions, changes in the depth of water, turbidity, tides and salinity regimes affect the performance and competitiveness and relationship of species.

Mangrove forest possesses the remarkable ability to regenerate when clear felled for timber and for other purposes. In the Wouri Estuary of

Cameroon Ndenecho (2007) observed that the mangrove plant recovery is very slow due to pollution problems. Pollutants originate from Urban and industrial waste and from the use of agro-chemicals in adjoining tree crop plantations. The pollution problems in the mangrove ecosystem in Cameroon are reviewed in the following paragraphs.

A great variety of pollutants are produced by man and many of these reach the aquatic environment. The sources of pollution are industrial and agricultural (agro-industrial, municipal and others) municipal sewage is the largest source of both nutrients and organic load. Tables 2 and 3 present the human activities along the coast that have polluting effects on the mangrove ecosystem. An investigation into the fisheries and levels of chemical pollution by Oben and Oben (2001) used plankton as bio-indicators of pollution. Copper values range from 0.002 – 0.24 ppm, well above allowed maxima in the European Union, Canada and USSR (0.005 –0.011mg/1), and are known to be toxic to some phytoplankton (Oben and Oben 1999). Increased nutrient inputs result from runoff from towns, industrial zones. Agro-industrial plantations and sewage discharge (Table 2) Nitrate values are also above the maximum allowable concentrations (40mg/ml). The low-species diversity in the Douala lagoon, Modeka, Tiko Creeks and Idenau River mouth suggest a stressed ecosystem, and the immediate vicinity of Douala industrial zone (Douala lagoon) has a high degree of pollution (Table 3)

Table 2: *Sources of pollutants into the Limbe Estuary*

Main sources	*Pollutants*
Industrial	Large or recalcitrant waste chemicals
Agriculture (Agro-industrial)	Pesticides and biocides, Herbicides, fungicides, Chemical fertilizers, Phosphates, Nitrates, Ammonium
Municipal	Landfill Domestic water & sewage Urban run off
Other	Silt Oil spills and leaks Oil from garages

(Source: Oben and Oben 2001)

An important source of pollution in the region is wastes from agricultural products, such as natural rubber (Tiko, Kukonje, Penda-boke, Nyiete factories), and palm oil mills (Idenau Modoni, Kienke). Palm oil and rubber are important because copious amounts are produced. Effluents from these factories are discharged into rivers and estuaries. Palm oil effluent is mainly inorganic and has a high biochemical oxygen demand (BOD) of 20.500mg/L (at 20^{0}C for 5 days). Although the effluent is biodegradable, its high BOD and suspended solids usually cause pollution when discharged into mangrove estuaries. Seow (1976) examined the effect of palm oil effluent on a mangrove in Indonesia's Sungei Puloh Estuary over a distance of 5km and in relation to tides. The BOD at the outfall exceeded 50mg/L at times, with maximum of 148mg/L during a neap tide, when effluent discharge was high, and even 5km upstream the BOD was 84mg/L. Mangrove within the immediate vicinity of the outfall appeared normal, but the mangrove fauna was depleted or absent (cf. Tables 2 and 3).

Table 3 Main activities along the Douala Lagoon and effects on the mangrove ecosystem

Activity	*Pollutants*	*Impact on the environment*
Agro-industry Food processing	Nitrates, nitrites, Chlorine, Phosphates, Organic Substances	Eutrophication Decrease in yield of food chain
Petroleum refineries Petroleum exploitation	Hydrocarbon, lead	Accumulation of lead and mercury in food chain Increased mortality of fauna and flora. Decrease in fish catch
Textiles Cement production Paint production Power generation Paper pulp production Soap production Bauxite processing	Acids, Mercury, lead, copper Phosphates, asbestos, organic substances	Accumulation of lead in food chain thereby decreasing yields
Battery production Engine oil production	Lead	Accumulation of lead in food chain

Port works	Sediments	Increase in turbidity, sedimentation and BOD
Production of plastics	Basic monomers	Accumulation of non-biodegradable material
Reclamation of wetlands	Solid wastes and sediments	Loss of habitat
Domestic activities	Human waste and organic substances	Eutrophication and decrease in oxygen supply
Building construction	Solid waste and sediments	Soil erosion Loss of vegetation and habitat
Fishing	Lead, lost nets	Destruction of mangroves and fish nurseries

(Source: Ibe *et al.,* 1999)

Sasekemer (1978) noted the ability of mangrove to tolerate palm oil effluent with a BOD several times higher than sewage; while Odum and Johannes (1975) found that mangrove ecosystem possess characteristics which facilitate their use for tertiary sewage treatment in the tropics. However Sasekemar highlighted the danger of dumping sewage among mangroves because of pesticides contamination from organochlorine, leading to increased pesticides residue in mangrove fauna, coastal fish and shellfish.

The main products of the oil refinery in Limbe include aviation fuel, petrol, kerosene, Diesel, Lubricants, wax, chemicals and bitumen. Probable pollutants include oil, arsenic, cadmium, hydrocarbons, chromium, copper, lead, nickel, Zinc and vanadium. Oil pollution in Limbe, Douala Port and the industrial zone is a constant threat to mangrove communities (Oben and Oben 2001). Spillages have been observed in shipping lanes and at the refinery. No information is yet available on the damage to the mangrove communities. Rutzler and Sterrer (1970) reported that mangrove communities on the Panama coast were seriously damaged by an oil spill; Rhizophora seedlings were killed as were inter-tidal invertebrates and algal mats. Mathias (1976) studied seedlings of the pioneer mangrove, *Avicennia Intermedia*, using diesel oil mixed with seawater in various ratios and found the concentration of 10.000 ppm and above killed seedlings.

The zones fringing the mangrove ecosystem are a microcosm of tropical tree crop plantation agriculture involving heavy dependence on herbicides.

Forest clearance in some cases involves initial blanket spraying of the forest to kill the trees before felling commences. In most cases, neighbouring mangrove forests are affected. Ross (1974) found that mangroves are very susceptible to herbicides, and areas affected by herbicides are resistant to recolonization, probably due to large quantities of wood debris, which restricts seedling germination and implantation.

From the silvicultural point of view, the mangrove swamps are the easiest to manage since seed planting is easy if natural regeneration is delayed (Dixon, 1959). Stands of *Rhizophora spp* are planted by sticking fruits into the mud. These take roots rapidly. On the single plant level, climate change will affect the mangrove phenology, that is, the time of the year when it starts flowering to bear fruits. The increased variability of precipitation in tropical regions might decrease the number of plant generations per year. In the tropics increasing temperatures are combined with a decreasing or at least more variable supply of water. This will adversely affect mangrove productivity.

Conclusion

Mangrove forests constitute a pan tropical ecosystem that has been considerably misused and mismanaged. The challenge to mangrove biologists is how to maximize the tolerance of the plant to pollution problems and to climate change impacts. Climate change has a major impact on the mangrove ecosystem, and therefore on the ecosystem services it provides. At the same time biodiversity in the form of the mangrove forest can play an important role in mitigating climate change through Carbon sequestration, coastal zone protection against tides and tsunamis, accretion at the coast and buffering the influence of climate change as well as supporting societies in adapting to it. Strategies to mitigate pollution problems, unsustainable exploitation and climate change impacts include:

- Institution and enforcement of laws protecting mangrove ecosystem
- Establishment of national mangrove sanctuaries to protect vital habitats and regenerate degraded areas.
- Rationalize industrial, agricultural and urban developments in mangrove zones.

- Use participatory approaches in adjoining communities to devise plans for the sustainable use and management of the mangrove ecosystem.
- Integrated coastal zone planning and management of coastal living resources must be an integral part of all development actions in coastal regions.
- Promote biodiversity governance at a global scale. For development cooperation to be sustainable under future climate change conditions there is the need for both the North and the South to reduce the emission of green house gases. The protection of the mangrove ecosystem is a global task; it cannot be regarded as the responsibility of the tropical and sub-tropical countries concerned.

References

Barbosa, L. A. (1970) Carta fitogeographica de Angola Luanada, Inst. Inves. Client Angola

Champman, V.J (1977) Wet Coastal ecosystems. Elsevier Scientific, Amsterdam.

Champman, V.J. (1971) Mangroves versus tidal waves. *Biol. Conservation*, Vol 4, p.39 -47

Cole, N.H (1968) The vegetation of Sierra Leone. Njalla University College Press, Njalla.

Dixon, F.G. (1959) A working plan for the Matang mangrove forest reserve, forest Department. Perak, Malaysia, 173p

Fondo, S (2001) The Socio-economic impacts of potential climate change on the Cameroon Coastal Zone. In: C.M. Lambi (ed) *Environmental Issues: Problems and prospects*. Unique Printers, Bamenda, p. 159-173

Fosesberg, F.R. (1971) Mangroves versus tidal waves *Biol. Conservation, Vol. 4* p. 38 - 9

Giglioli, M.E. and Thornton, I. (1965) The contribution of Mangrove swamps of Keneba, lower Gambia River Basin. *Journal of Applied Ecology,* Vol. 2 No 1, p. 81-103

Heald, J.H. and Odum, W.E. (1969) The contribution of mangrove swamps to Florida fisheries. *Proc. Gulf Carrib Fish. Inst.,* Vol. 22, p. 130 – 5

Kuete, M. (1998) Le milieu de L'ecosyteme mangrove: le cas de bouches du Cameroun *Journal of applied Social Sciences,* Vol. 1, P. 91-109

Longman, K.A. and Jenik, J. (1987) Tropical forest and its environment Longman Scientific and Technical, 347p.

Mathias, J.A. (1976) The effect of oil on seedlings of the pioneer Mangrove, *Avicennia intermedia,* in Malaysia. *Abstracts of Symposia and contributed Papers,* International Symposia on the Ecology and management of some Tropical Shallow water Communities, 27 June – 10 July 1976, Jakarta.

Millington, A.C. (1987) Environmental degradation, soil conservation and agricultural policies in Sierra Leone, 1895 – 1984. In D. David and R. Grove (eds) *Conservation in Africa people: policies and Practice.* Cambridge University Press.

Millington, A.C, Critchley, R.W., Douglas, T.D. and Ryan, P (1994) Estimating woody biomass in sub Saharan Africa. The World Bank, Washington D.C, 191p

Ndenecho, E.N. (2008) Economic value and management of mangrove forests in Cameroon. Int. *Journal of sustainable Development and World Ecology,* Vol. 14, p. 618 – 625

Oben, P.M. and Oben, B.O. (1999) Plankton Survey. Final Report on Plankton survey in natural waters. *UNDP Large Marine Ecosystem Project, 180 p*

Oben, P.M and Oben, B.O (2001) A review of the constraints to the development of the fishery sector in Cameroon. In: Lambi C.M (eds.) *Environmental issues: problems and prospects*, Unique Printers, Bamenda p. 67-84

Odum, W. E. (1970) Pathways of energy flow in a south Florida estuary. Dissertation, University of Miami, 180 p

Odum, W.E. and Johannes (1975) The response of mangroves to man – induced environmental stress. In E.J Wood and R.E Johannes (eds) *Tropical Marine Pollution*, Elsevier Scientific Publishing Company, Amsterdam, p. 52 – 62

Pathansali, D (1976) Assessment of potential yields from the coastal marine fisheries resources of Malaysia. *Fish Bull Ministry of Agriculture,* Vol. 15 59p,

Peare, D.W and Warford, J.J. (1994) World without end. Economics, Environment and sustainable Development O.U.P, New York.

Pepper, W. (1992) Emission scenarios for IPCC. An update. *Intergovernmental Panel on the climate change working Group,* IPCC.

Ross, P. (1974) The effects of herbicides on the mangrove of South Vietnam (*The Effects of Herbicides in South Vietnam, Part B, Working Papers.* National Academy of Sciences, Washington D.C., P1-33

Rutzler, K. and Sterrer, W. (1970) Oil pollution damage observed in tropical communities along the Atlantic Seaboard of Panama. *Bioscience,* Vol. 20, P. 222 - 4

Sasekemar, S. (1978) The Value of the Mangrove Ecosystem and its pollution problems. 5th FAO/SIDA Workshop. *Aquatic pollution in relation to protection of living resources* F.A.O Rome, p. 332 – 342.

Seow, R.C.W. (1976) The effect of a mixed organic effluent on the distribution of pelagic macro-fauna at Sungei Puloh with special reference to water quality. BSc. Thesis, University of Malaysia, 43p.

Spanniks, F and Van Beukering, P, (1979) Economic valuation of mangrove ecosystems: Potential and limitations. *CREED working Paper* No 14. 11 ED London

Vohland, K. (2008) Biodiversity and climate change. *The inter. Journal for Rural Development*, Vol, 42, No .2

Youmbi, T., Angwe, C. and Ndongo D. (1997) Assessment of climate change impacts on the Cameroon Estuary mangrove ecosystem. MINEF/UNDP, Yaounde.

Chapter Four

Climate Change and Sudan-Sahelian Savannah Ecosystem Managment

Summary

It is now abundantly clear that global climates are changing and ecosystems are vulnerable to those changes. Sub-Saharan Africa according to experts will experience increased temperatures and decreased precipitation. The Savanna biome will be altered by these changes. There is therefore a need to protect and enhance its resilience to climate change. The chapter overviews the climate change scenarios and range vulnerability in the Sudano-sahelian savannah of Africa using Cameroon as a case study. It concludes that despite indigenous adaptations to climate change by pastoral tribes, public policy and awareness in conserving and enhancing the resilience of the ecosystem is lacking. It therefore overviews adaptation strategies that can direct decisions to cause a public awareness and assist policymakers to implement policies and educational measures that can counter the threat. Finally, it provides policy guidelines that can assist the governments of sub-Saharan African countries adjust livestock economies to climate change impacts in the savannah biome.

Key Words: Climate change, sudano-sahelian savannah, enhancing resilience, management policy, political reaction, public awareness.

Introduction

Available literature makes it abundantly clear that the climate is changing and ecosystems are being affected by these changes. Much awareness has been raised about the effects of climate change on natural systems. The structure and function of the savannah biome of Africa makes it vulnerable to global climate change and human environmental impacts. The effects of climate change are now visible in grasslands around the world. Parmesan and Yohe (2003) analyzed data on more than 1700 species to show with a "very high confidence" level that climate change has already altered range boundaries and phenology. A separate study using data from 143 studies found that 80% of species studied showed trait changes consistent with climate change - driven predictions (Root *et al.*, 2003). This chapter overviews the climate change characteristics, and range vulnerability in the Sudano-Sahelian region of Africa using Cameroon as a case study. It seeks to devise adaptation strategies that can direct decisions to maximize resilience of the Savannas to climate change. The objective is to cause a public awareness and to assist policy makers to respond to this growing threat.

Climate Change Characteristics

The climate is mainly sudano-sahelian, and it is characterized by wet and dry seasons. The mean annual rainfall amount varies between 600-1,000 mm while the number of dry months with a monthly rainfall of less than 50mm ranges from 6 to 7 months. Climatic variations of a short-term nature are superimposed on the longer period climatic fluctuations of the Quaternary. These have affected the history of plant communities.

Table 1: *Frequency of variability of Annual Rainfall (1970-1988)*

Probability of rainfall greater than the threshold indicated	Threshold rainfall (mm)			
	Kaele	Maroua	Garoua	Touboura
0.8	622	631	820	987
0.5	655	675	883	1054
0.2	781	780	1014	1247
Average for the period	666	752	951	1180

Source: Donfack *et al.*, 1996

From table 1 it can be observed that inter-annual rainfall variability is typical of the Sudano-Sahelian area. Table 1 represents the frequency of variability of rainfall between 1970 and 1988. These variations influence the cycles of biological activity and therefore a rhythm of feasts and famines for cattle. During drought years livestock experience a drop in weight. The droughts of the 1970s were characterized by the loss of livestock and human life.

Range Resources Vulnerability to Climate Change

More than 7 million hectares of rangeland are in the area. The South Benue area is restricted by the prevalence of the tsetse fly. The native vegetation in the northern parts of the study area has been greatly depleted by many years of burning, overgrazing and other detrimental practices. Much of this area was once perennial grassland and open savannah. It is now covered with annual grasses, weeds and unpalatable shrubs, and the amount of forage presently produced is less than half of that originally produced.

The consequences of this climatic situation are a biological rhythm with two phases:

- An active vegetative phase of very intense growth of about 5 months. The reconstitution of ground water reserves is essentially achieved from July to August when rainfall exceeds evapotranspiration. This reserve is rapidly depleted as soon as the dry season sets in.
- The wet season is followed by a dry hot season during which ground water reserves are exhausted by high rates of evaporation. The physiological activities of plants are very restricted and much of the vegetation dries out.

The overgrazing of upland pastures during the dry season degrades these sites. The growing season is limited. Together with frequent droughts the risk of famine is high and there is a legacy of vegetation degradation. Table 2 and figure 1 present the natural vegetation communities and their yields.

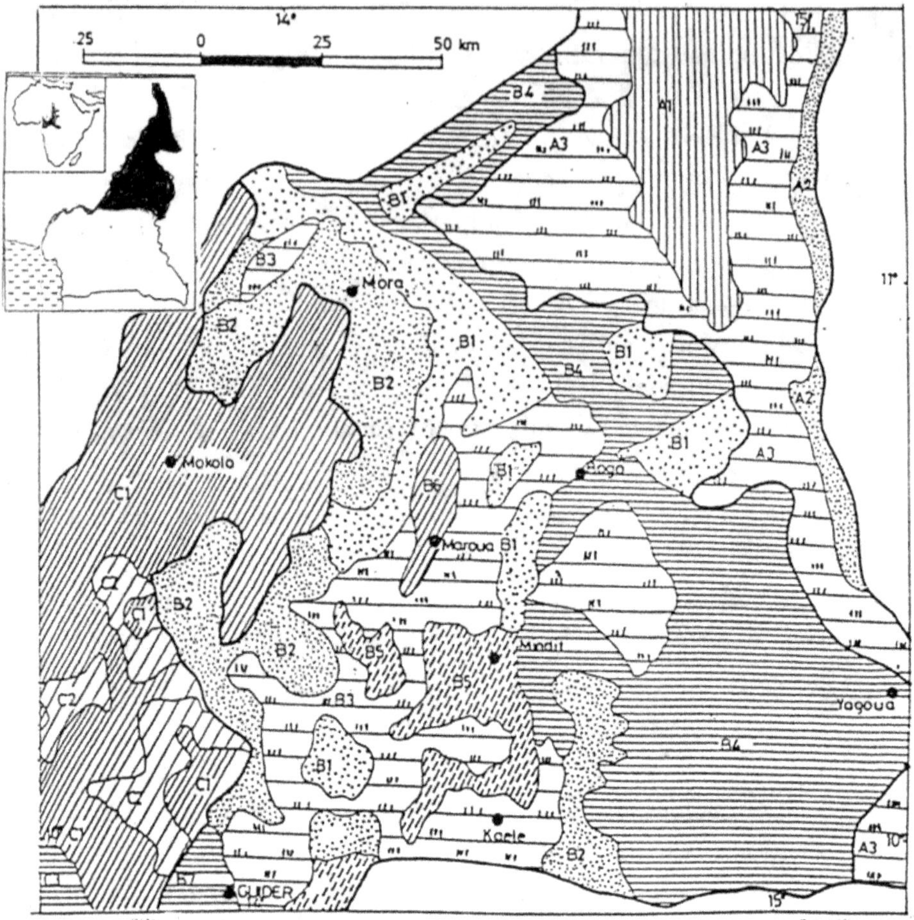

Figure 1: Morphologic Units and natural plant Communities of North Cameroon.

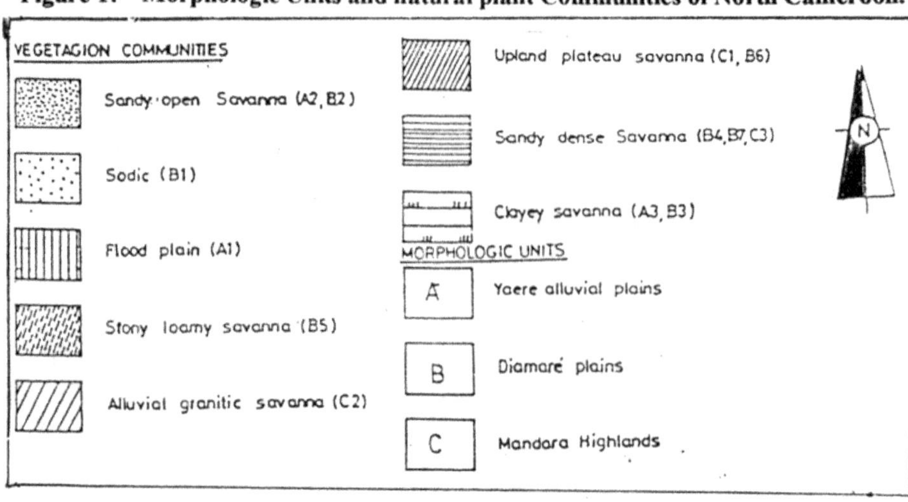

Table 2: *Natural plant Communities in the Sudano-Sahelian region of Cameroon and estimated yields*

Name and map code (Figure 1)	Composition	Estimated yield (kg/ha)
Sandy Open Savanna (A2, B2)	*Hyparrhenia rufa* *Pennisetum pedicellatum* *Ctenium species* *Aristida species* *Eragrostis tremula* *Acacia species* *Combretum giutinosa* Annual grasses	500-1000
Sodic (B1)	*Acacia species* Annual grasses	100-500
Flood plain (A1)	*Hyparrhenia rufa* *Sporobolus pyramidalis* *Setasia polidisfusea* *Rottobellia exaltata* Annual Sorghums	8000-10.000
Stony loamy savannah (B5)	*Hyparrhenia rufa* *Eragostis species* *Aristida species* *Acacia Senegal* *Balanites aegyptica*	600-1200
Alluvial granitic savannah (C2)	*Hyparrhenia rufa* *Pennietum pedicellatum* *Eragrostis robusta* Annual grasses *Ficus species* *Balanites aegyptica* *Acacia albida* *Acaci seyal* *Acacia Senegal* *Acacia tortilis* *Combretum glutinosum*	600-1200
Upland plateau (C1, B6)	*Hyparrhenia rufa* *Setaria palidisfusca*	1500-2500
Sandy dense savannah (B4, B7, C3)	*Hyparrhenia rufa* *Pennietum pedicellatum* *Andopogen gayanus* *Cymbopogon proximus* *Ctenium canesiens* *Aristida species*	3000-5000

	Commiphora African *Seleracarya bierre*	
Clayey savannah (A3, B3)	*Aristida species* *Pennietum pedicellatum* *Annual Setarias* Other annual grasses *Acacia species* *Combretum glutinosa*	800-1500

Source: Fulton *et al.,* 1974

Overgrazing is intense in the Diamare Plains where the land/cattle ratio is very high (2 head/ha). About 40% of the total cattle production in Cameroon is in these dry lands. In 1975, there were 58.100 herds of cattle, 698.000 goats, 580.000 sheep, 52.100 donkeys and 7.165 horses grazing on 2.208.100 hectares of rangeland in the Mandara-Diamare and Mayo Danai area. The consequences of such high animal densities are:

- a decline in annual production of pasture vegetation;
- a decline in the palatable grass species, particularly perennials which are also good at holding the soil together (soil aggregation)
- an increase in ephemeral plants, which spring up with the on set of the rains rather than having a permanent presence, thus decreasing the durability of pastures.
- Soil compaction as a result of sealing and trampling by stock near water holes and overgrazing wet-season pastures; and
- Damage to vegetation on crests and stable dunes resulting in erosion (erosion of palaeo-sand dunes).

Table 2 presents the degree of vegetation degradation per floristic community. A variable climate with recurrent droughts and annual burning of the natural pastures triggers a legacy of overgrazing and vegetation degradation. Climate change paves the way for the invasion of natural pastures by annuals and other unpalatable species.

Table 2: *Degree of vegetation degradation in the Sudano-Sahelian region .*

Land unit	Map code	Degree of Degradation of the Potential Vegetation community
Sodic	B1	75 to 100% of the potential plant community has been degraded and replaced by annuals.
Sandy open savannah	A2, B2	50 to 75 of the potential plant community has been degraded and replaced by annuals or other species
Alluvial granitic savannah	C2	
Clayey savannah	A3, B3	25 to 50% of the potential plant community has been degraded and replaced by annuals or other species.
Stony loamy savannah	B5	
Sandy dense savannah	B4, B7, C3	
Flood plain	A1	The present community is almost the same as the potential plant community. Degradation has been minimal.
Upland plateau	C1, B6	

Source: Modified after Fulton *et al.* 1974.

"Nomadic pastoralism is inherently self-destructive, since systems of management are based on the short-term objective of keeping as many animals as possible alive, without regard to the long term conservation of land resources" (Allan, 1976). Poor pasture management arising from archaic social customs whereby grazing rights are based on customs and history of land use, and the lack of boreholes for watering livestock during the long dry season, reinforce the constant movement of cattle. Much of this zone which serves as the transhumant corridors for cattle from Nigeria to Central African Republic or from Chad to Nigeria, help to aggravate the desertification problem. This zone has a total surface area of 10 million hectares (Fulton, 1974) which has been divided into different land use types. Of these, more than 7.5 million hectares representing 76.5% consist of rangelands, most of which are poorly managed and overgrazed. While only 6% of the land hectarage is devoted to agricultural production, 12% passes for game reserves; the other forms of land use account for the remaining 5.5%.

The low-structure of the savannah vegetation confers high light availability, leaving it naturally vulnerable to invasive species (Wagner, 1989), especially following disturbance (Hobbs and Huenneke, 1992) by grazing pressure and burning. A critical role for fire in maintaining the savannah leaves it vulnerable to vegetation change should changes in temperature and precipitation occur that are significant to alter biomass and fire frequency.

Global warming accompanied by droughts will increase fire frequency and reduce the biological cycles. There is therefore a need for scientists, policymakers and land managers to devise and implement adaptations that will be needed to protect and conserve the natural pasture from threats posed by global climate change.

The conditions of vegetation and soils will prove critical to the resilience of the savannah in this eco-climate. Healthy, rigorous stands of native vegetation are likely to be more resilient to warming temperatures and increasing frequency and duration of droughts than degraded grasslands because their roots have access to deeper moisture and they are better able to compete with invasive species (Godwin *et al.,* 1999; Enloe *et al.,* 2000; Gelbard, 2003). Healthy plant cover is crucial for intercepting rainfall, maximizing infiltration and soil water supply, reducing overland flow, and preventing nutrient losses due to erosion (Noss and Cooperrider, 1994). A healthy level of soil organic matter is important for soil aggregate formation, fertility, stability, water movement and holding capacity and aeration and therefore influences plant growth. Plant growth and vigour is largely determined by physical, biotic and disturbance factors that limit primary productivity, including water availability (Sala *et al.,* 1988), nutrient cycling (Evans and Belnap, 1999), the suite of competing species (Grace and Tilman, 1990), healthy soil biota and the native disturbance regime such as herbivory and fire (Collins, 1992). Today these conditions are being altered by a combination of anthropogenic and climate change factors:

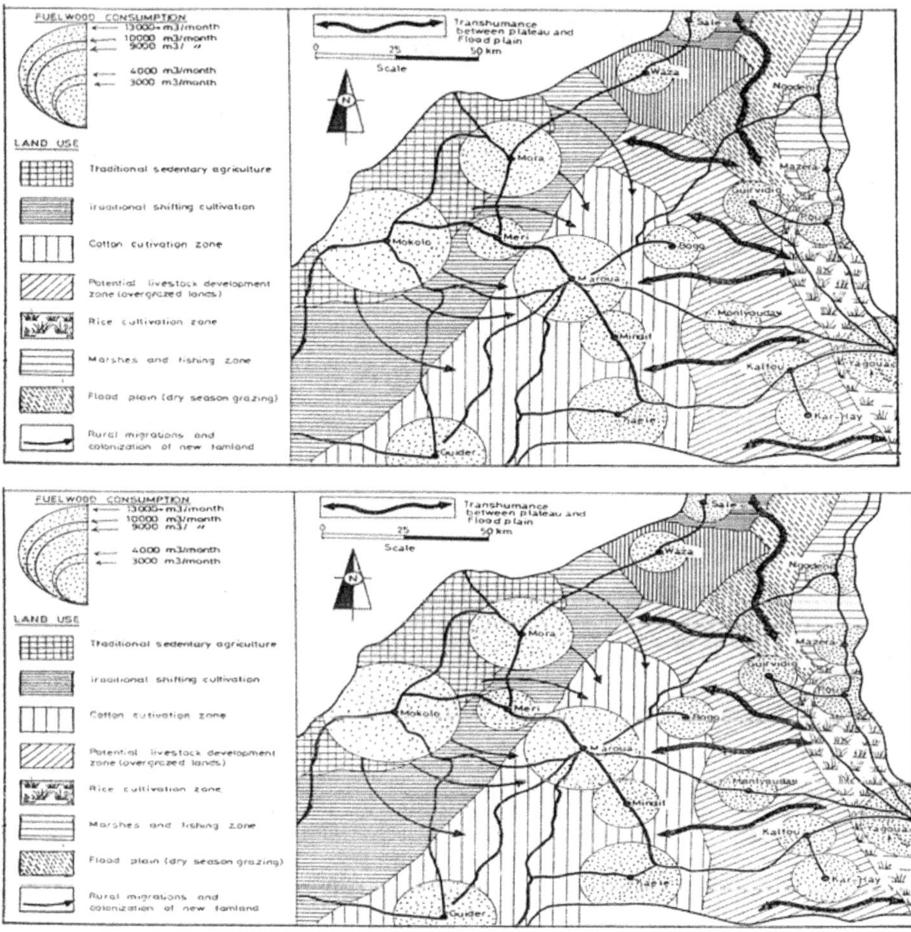

Figure 2: Vegetation exploitation in sudano-sahelian Cameroon

- Overgrazing by livestock: In most range sites 50% to 75% of the potential plant community has been degraded.
- Urbanization: Figure 2 presents the urban centres in the study area. These centres and their surrounding are vulnerable to habitat loss due to urbanization.
- Spread of invasive species: On most range sites 50 to 75% of the potential plant community have been degraded and replaced by annuals and other species.
- Altered fire regimes: Prolonged droughts resulting from climate variability alter fire regimes which are necessary in maintaining the savannas. Increased fire frequency will further degrade the steppe savannas.

Livestock Resources and Endogenous Climate Change Adaptations

There are more than a million head of cattle in the study area, or about 40% of the national total. Average cattle density is 1 animal for every 3.5 hectares. Average meat production is about 12kg per person. The kinds and breeds of domestic animals reflect the harsh environment, traditional values, and expressed socio-economic needs of the people. The livestock industry is at subsistence production level. Traditionally, the herdsmen, who either are nomads who keep cattle as a way of life or villagers who keep them as an adjunct to agriculture, raise livestock without significant capital investments except in their animals. Milk is basic to human survival, and beef from surplus animals is secondary. Sheep and goats generally are not raised to provide milk; instead they supply meat for consumption. Their economic importance is minor compared to that of cattle. Many people live at subsistence level on this relatively small per capita, number of animals, but they do so at the expense of a deteriorating land resource base.

The classification of cattle, sheep and goats by eco-type and tribal breed is a follows:

- Cattle: (a) Trypano-susceptible

  Zebu (humped), *Bos indicus*
    - Choa Arab or Wahdara
    - M'bororo Fulani or Rahaji
    - Fulbe

Taurine (humpless), *Bos taurus*

- Kurt, Buduma, or lake

(b) Trypano- tolerant

Taurine (humpless)

- Kapsiki
- Poli

- Sheep: (a) Deserto – sahel
    - zaghawa
    - ouda
    - Arab

  (b) Sudan-Guinea crossbred
    - chao Arab
    - Foulbe

  (c) Equatorial and Kirdi enclave
    - Kotoko
    - Rain forest dwarf

- Goats: (a) Deserto–sahel
    - zaghawa
    - Foulbe
    - Arab

  (b) Sudano-Guinea crossbred

Under existing baseline practices, breeding is not a major limitation to production among zonal livestock. Although some tribal breeds and some resource areas can be made more productive by greater attention to breeding, more favourable results can be directed at clearing areas of tsetse flies and introducing more productive cattle, upgrading with better bulls and encouraging immigration. More southerly tribal breed sheep can be upgraded with larger Deserto-sahel rams as a measure to combat climate change. Goats live in an environment that reflects poverty and degraded land resource base. Their environment favours only survival of the fittest, natural selection, therefore, has produced the best possible gene combination.

Under climate change conditions characterized by long droughts nutrition is the main limitation to baseline livestock production regardless of species and resource area. Traditional producers have long countered feed shortages during the long dry season by the use of migratory and semi-sedentary husbandry practices. However, the animal energy required to reach food and water and the dependence of producers on milk during seasonal migration are serious limitations to productivity. Also, the starving of calves from lack of milk precludes the attainment of production levels beyond a certain point. The production of goats is limited by poor nutrition than that of cattle or sheep. Producers react to fluctuations in pasture ecological resources by adopting opportunistic grazing systems (Beauvilain, 1981):

- Following the rainfall within the region: Incidence and quantity of rainfall vary, not only from year to year but also spatially. Herd movement tends be from drier to wetter areas within a region. In this way forage supply can be balanced to some extent.

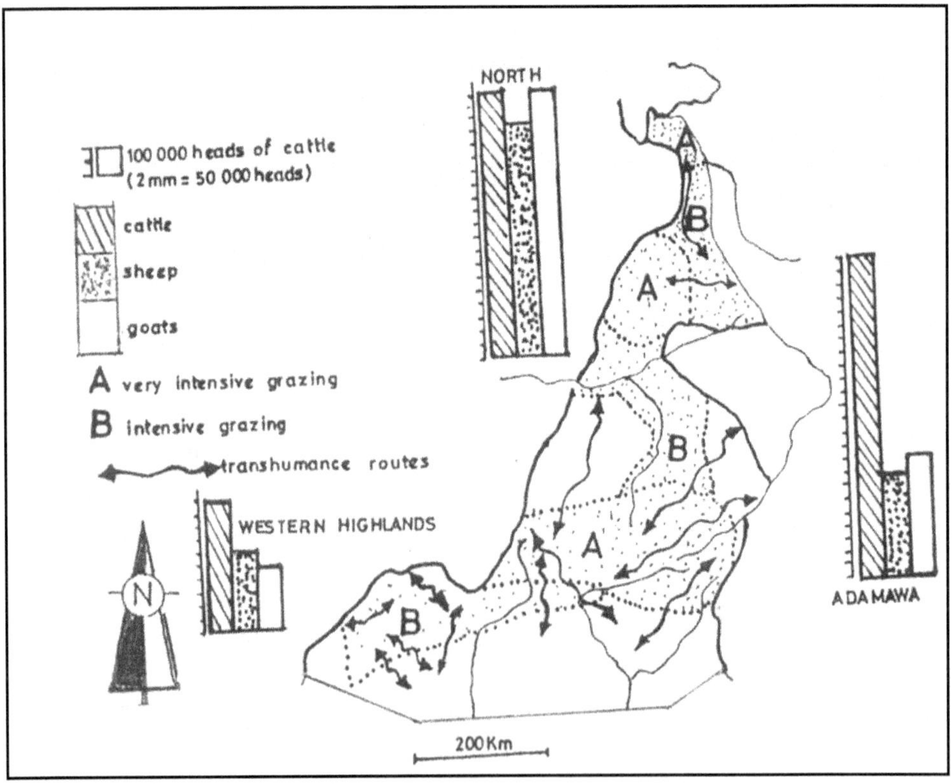

Figure 3: Livestock production and adaptation of grazing to climate variability in the sudano-sahelian Cameroon

- Securing access to key resources: If forage is scarce in the "normal" pastures herdsmen move cattle to reserve pastures in flood plain and plain zones which receive storm water and floods. The thorny issue of land tenure makes this access difficult.

- Moving between different agro-ecological zones: After the harvest of millet, sorghum, groundnuts and rice, the stubble offers decent fodder during the dry season. Again land tenure makes this access difficult.

Endogenous developments in the Sudano-Sahelian region of Cameroon following the droughts of the 1970's and 1980's indicate that herders are switching species, diversifying their portfolio of animals and are moving away from cattle and towards sheep. They are expected to make these changes without much change in expected income. Policy makers in the country must be careful to encourage private adaptation during this period of change. There may be nothing that can be done to sustain large cattle herders that depend on current climate. Providing subsidies and other enticements for such operations to continue once the climate changes would only compound the problem. Instead, the government should encourage farmers to change the composition of animals on their farms as needed. There is a need to share indigenous knowledge on how livestock owners have coped with higher temperatures. The government should anticipate that farmers will make changes on their lands and do whatever is needed to facilitate these changes. Some indigenous adaptations include: herd supplementation, herd movement following rainfall opportunistic events and edaphic floristic communities, livestock diversification, culling animals and reduction of herd / farm sizes.

In spite of the vantage point of climate change in the international politics and media the political will to counter climate change in Cameroon is only coming up now. For measures to be successful future climate change condition adaptations must be an intrinsic part of international and national legislation as well as government and administration. Current institutions that are supposed to tackle this challenge posed by climate change do not have reliable programmes. There is need for a coordinated programme of action. Sustainable adaptation will hardly be achieved without improved human and institutional capacity building. There is the need to develop partnerships with global institutions that possess resources and lessons for climate adaptation strategies.

- Protect representative grassland types per topographic unit and floristic community. Because we do not know precisely which plant community will be most sensitive to climate change, maintaining grazing reserves in replicated protected areas per ecological niche to protect against the loss of individuals to catastrophic events is necessary. Appropriate management to native disturbances regime can ensure resilience. Such reserve systems should include heterogeneous topographic sites, soil and management conditions to maintain native species.
- Protect relict, native-dominated plant communities: These are found in remote topographic sites. These are refugia which can serve as models for habitat restoration. They can help us understand how altered and unaltered grasslands by human activity are affected by climate change (Ambos *et al.*, 2000).
- Minimize the fragmentation of natural pastures by other land use systems: Land use systems such as agriculture, urbanization and road construction fragment the climatic climax communities making them susceptible to invasion by undesirable species (Dale *et al.*, 2000).
- Promote low-intensity sustainable grazing practices: Low-intensity grazing should be promoted in plant communities where native species are adapted to it in order to encourage resilience (Mack, 1989). It should be discouraged on marginal lands where the dominant native species lack a long evolutionary history of grazing adaptation (Milchunas and Lauenroth, 1993).
- Prevent and control the spread of invasive species and pathogens: since invasive species may be favoured over natives under climate change conditions, and cause degradation to native species, preventing and controlling their spread is critical for maintaining resilience (Mooney and Hobbs, 2000). For example *Acacia Spp.* now dominates many sites. To restore grazing use of these areas, grazing can be deferred until annuals and perennial grasses can become established. *Acacia Spp.* can then be controlled by burning the sites and hand clearing the *Acacia.* When perennial grasses are re-established, they compete with the *Acacia* and help to control it. This

should be part of a grassland restoration programme involving the reintroduction of native species, and inoculation with soil biota important to native plant regimes (Soul and Terbourgh, 1999). The Logone-Chari Dalta and Diamare Plains are badly denuded and are possible areas for establishing grazing reserves. The government can select the area, plan the development, stock the reserve with livestock, and oversee the entire operation. Alternatively, it can give professional assistance to tribal groups in developing and managing such sites. The latter approach offers the most promise.

- Promote the maintenance of natural fire regimes: Many areas in the Logone-Chari Delta and Diamare Plains are nearly barren because of extreme grazing pressures and frequent fires. *Acacia spp.* now dominates these sites. Fire regimes exert a profound influence on the health and heterogeneity of grassland vegetation (Ricketts and Dinerstein, 2001). Maintaining or restoring native fire regimes can therefore restore grassland resilience (Dale *et al.*, 2000).
- Identify and protect keystone species: The steppe savannah are dominated by *Hyparrhenia spp., Pennisetum pedicellatum, Eragrostis spp., Setaria spp.* and *Aristida spp.* Maintaining the natural diversity of species and functional groups, and dominance of native species, is a sound overall strategy for enhancing resilience to climate change (Dukes, 2002).

The feasibility of implementing the above climate change adaptation strategies will require the following strategies:

- Landscape level land use planning with the objective of maintaining and restoring resilience. This will require conservation-oriented planning for local decision-making at regional level (Dale *et al.*, 2000)
- Develop programmes to slow and prevent the proliferation of invasive species. Strategies to achieve this goal should focus on the causes of invasions such as seed sources and disturbances that increase vulnerability to invasions (Sheley and Petroff, 1999).
- Implement science-based and adaptive livestock management strategies. Design appropriate grazing systems and management on a site-specific basis depending on soil, soil moisture, predominant

native species, life form and invasive species that pose a threat (Holecheck *et al.,* 1998).

- Undertake grassland restoration programmes on depleted sites. Such restoration actions will depend on the factors reducing resilience and are therefore site-specific. In the face of climate change actions are needed to maintain the composition and structure of plant communities.

Action and Policy Framework

Policy and management actions required to implement the above climate change adaptation strategies can be implemented at the regional, national, community and individual levels. At the regional level, they involve the control of the movement of cattle from Nigeria, Niger, Chad and the Central African Republic to dry season pastures in the Logone-Chari flood plain and Delta. At the national and local levels, appropriate measures will differ between public and private grasslands, and perhaps among tribal groups.

The implementation of adaptation strategies will be easier if ranches are organized and if tribal leaders are cooperative and interested. Development costs for such ranches are lower, because the tribe or group can voluntarily contribute to grassland management. Furthermore, labour costs are reduced, and no special marketing arrangements are required. They will use traditional market channels. Besides being more economical to establish than government –operated ranches, tribal-operated ranges will encourage tribal unity. Family groups and tribal leaders can work together to improve their livelihoods in a climate change scenario rather than have an outside supervisory structure imposed on them. Tribal-operated ranches encourage the fullest use of the country's resources, that is, people, herds, land, water. Such ranches can be expanded with relatively few social and economic constraints.

Professional assistance would be supplied by the government. The government would work to educate livestock owners and to develop among them an awareness to climate change vulnerability and the benefits of the implementation of adaptation measures. On the other hand, it would be the responsibility of tribes to enforce grazing plans and to implement science - based livestock management and grassland management strategies.

An essential part of this plan will be the selection of homogenous tribal groups with recognized leaders who are willing to associate themselves with plans to manage their traditional grazing lands. Range size will depend on the nature of tribal group, the extent of their respected leadership, the productive capacity of the grassland, and to some extent topography.

Some large ranching enterprises undertaken in Africa have failed. It is advisable, therefore, to start with a small operation, and as the programme becomes successful, the operation can be enlarged. Problems which evolve in a new enterprise can more easily be resolved successfully for a small operation than for a large one. Whichever approach is used, it is essential to carefully evaluate all the resources, and develop a simple but sustainable management plan. Each plan should fully accommodate the climate change adaptive strategies and the needs of the native plants.

These tribal groups and private landowners will need financial impetus and know-how to implement climate change adaptations. It will be important for governmental agencies and non-governmental organizations (NGOs) to use incentives to convey a sense that changing management will provide an opportunity to improve economic well-being and quality of life.

It will also be important to educate the public about the availability of incentive programmes. Both NGOs and governmental agencies will prove key as grassland owners seek assurances that taking steps to increase resilience will not threaten their livelihoods. Financial and quality of life concerns must be addressed by providing both direct (conservation easements, conservation rewards, habitat conservation plans, grants, cost-share programmes) and indirect (education, technical support) benefits to make grassland conservation and restoration economically and socially desirable, and thus more politically feasible (McNeely, 2000; Naylor, 2000).

Conclusion

Implementing the proposed adaptations will undoubtedly pose considerable challenges. The most effective options for maximizing resilience may not always appear socio-economically feasible, and may meet with considerable political resistance. Ecologist, and land managers will therefore need to work together with economists, sociologists, communication specialists, policymakers and the public to devise economic and public relation strategies aimed at gaining the support of sceptical policymakers, tribal groups and landowners. Funding for devising and implementing adaptations is a critical factor. This may come from international organizations such as the United Nations Environment Programme (UNEP), International Union for the Conservation of Nature (IUCN), the World Commission on Protected Areas, and the Global Environment Facility. The Ministry of Forests and the Environment, Ministry of Agriculture, and the Ministry of Livestock and Animal Husbandry may also provide financial support. Technical expertise for implementing adaptations could be provided by NGOs, university experts, the Zootechnical Research Institute experts, graduate students and the extension service of the Ministry of Livestock and Animal Husbandry. There will be a need for human capacity building for range management experts and technician. This can be achieved by:

- sending qualified personnel to countries that have range management schools and on-location training programmes
- establishing management courses in the local university system
- establishing training locations in the sudano-sahelian region staffed by qualified grassland managers
- introducing range management courses at centres for training farm families.

References

Allan, W. (1976) The African husbandman. Oliver and Boyd, Edinburgh. 321p.

Ambas, N.; Robertson, G. and Gouglas, J. (2000) Dutchwoman Butte: a relict grassland in Central Arizona. *Rangelands Vol. 22,* p. 22-24

Ayonghe, S (2001) A quantitative evaluation of global warming and precipitation in Cameroon from 1930 to 1995 and projections to 2060: effects on the environment and water resources. In: C.M. Lambi and E.B. Eze (eds.) *Readings in Geography,* Unique Printers, Bamenda. p. 142-155.

Beauvilain, A. (1981) "Elevage et eleveurs dans le Grand Yaeé-Nord Cameroun" *Cameroon Geographical Review, Vol. 2, No.2,* University of Yaounde, p. 163-176.

Carter, R. and Alkali, A. (1996) Shallow groundwater in the northeastern arid zone of Nigeria. *Quarterly Journal of Engineering Geology Vol. 29,* p. 341-355.

Collins, S. (1992) fire frequency and community heterogeneity in tall grass prairies vegetation. *Ecology Vol. 73*, p. 2001-2006.

Dale, V. Brown, S.; Haeuber, N; Hobbs, T.; Hunty, R.; Naiman, J.; Riebsame, W.; Turner, M.; and Valone, T. (2000) Ecological principles and guidelines for managing the use of land. *Ecological applications vol. 10,* p. 639-670.

Donfack, P.; Bouker, S.; M'Biandown, M. (1996) Les grandes characteristiques du milieu physique. In: Agricultures des savannas du Cameroun. Actes de l'atelier d'echange, CIRAD, Maroua. p. 29-30.

Dukes, J. (2002) species composition and diversity affect grassland susceptibility and response to invasion. *Ecological Applications Vol. 12,* p. 602-617.

Enloe, S.; Di'Tomaso, J.; Osloff, S. and Drake, D. (2000) integrated strategies for the attrition of yellow starthistle on Northern California rangelands. *Proceedings: California Weed Science Society Vol. 52,* p. 31-34.

Evans, R. and Belnap, J. (1999) Long-term consequences of disturbance on nitrogen dynamics in an arid ecosysytem. *Ecology Vol. 80,* p. 150-160.

Fulton, H. Donald, Landry, P.; Bower, J. and Boulet, J. (1974). Resource Inventory of North Cameroon, United states Department of Agriculture, Soil Conservation Service, USAID-FAC-France, p. 64-67.

Gelbard, J. (2003) Understanding the distribution of native versus exotic plant diversity in California's grassland landscapes. Ph.D. Dissertation, University of California at Davis.

Goodwin, J.; Doescher, L.; Eddleman, L.; and Zabel, D. (1999) Persistence of *Idaho fescue* on degraded sagebrush steppe. *Journal of Range Management Vol. 52,* p. 187-198.

Grace, J. and Tilman, D. (1990) Perspectives on plant competition. Academic Press, New York. 484p.

Hobbs, R. and Huenneke, L. (1992) Disturbance, diversity and invasion: implications for conservation. *Conservation Biology vol. 6*, p. 324-337.

Helecheck, J.; Pieper, R. and Herbel, C. (1998) Range management principles and practices. 3rd Edition, Prentice Hall, Upper Saddle River.

Mack, R. (1989) Temperate grasslands vulnerable to plant invasions: characteristics and consequences. In: J. Drake, h. Mooney, F. Di Castri, R. Groves, F. Kruger, M. Rejmanek and M. William (eds.) *Ecological invasions: a global perspective.* John Wiley and Sons. Chinchester. 525p.

McNeely, J. (2000) The future of alien invasion species: changing social view. In: H. Mooney and R. Hobbs (eds.) *Invasive species in a changing world.* Island Press, Washington D.C. p. 171-190

Milchunas, D. and Lauenroth, W. (1993) Quantitative effects of grazing on vegetation and soils over a global range of environments. *Biological Monographs No. 63,* p. 327-366.

Mooney, H. and Hobbs, R. (2000) Invasive species in a changing world. Island Press, Washington D.C. 457p.

Naylor, R. (2000) The economies of alien species invasions. In: H. Money and R. Hobbs (eds.) *Invasive species in a changing world.* Island Press, Washington D.C. p. 241-260.

Noss, R. and Cooperrider, A. (1994) Managing rangelands. Chapter 7, In: *Saving Nature's Legacy: protecting and restoring biodiversity.* Island Press, Washington D.C.

Parmesan, C. and Yohe, G. (2003) Global climate change and biodiversity conference presentation. University of East Anglia, Norwich, 8-10 April, 2003.

Ricketts, T. and Dinerstein, E. (2001) Terrestrial ecoregions of North America: a conservation assessment. Island Press Washington D.C.

Root, T.; Price, J.; Hall, K.; Schneider, S.; Rosenzweig. C.; and Pounds, J. (2003) Finger prints of global warming on wild animals and plants. *Nature Vol. 431*, p. 57-60

Sola, O.; Parton, W.; Joyce, L.; and Lawenroth, W. (1988) Primary production of the central grassland region of the United States. *Ecology Vol. 69*, p. 40-45.

Sheley, R. and Petroff, J. (1999) Biology and management of noxious rangeland weeds. Oregon State University Press, Corvallis.

Soule, M. and Terbourgh, J. (1999) continental conservation. Island Press, Washington D.C.

Wagner, F. (1989) Grazers, past and present. In: L. Huenneke and H. Mooney (eds.) *Grassland Structure and function: California annual grassland.* Kluwer Academic Publisher, Dordrect, The Netherlands.

Chapter Five

Climate Change and Montane Biodiversity Management

Summary

The montane forest of Cameroon like most afromontane forest refugia in Africa are unique. They have one of the highest levels of endemism particularly among birds and vascular plants. Biodiversity is extraordinarily rich. Despite their scientific importance they have received little conservation attention and so have suffered from degradation. The chapter examines the biodiversity features, makes observations on current status of the forests, the types and severity of threats and establishes how climate change affects biodiversity. It concludes that climate change has major adverse impact on biodiversity, and therefore also on ecosystem services such as food and water availability. The conservation and diversification of genetic resources can play a fundamental role in mitigating climate change, buffering the influence of climate change as well as supporting rural communities of sub-Saharan Africa in adapting to it. It finally identifies the scope for adapting biodiversity management, food security livelihood sustenance and poverty alleviation strategies to climate change.

Key Words: Climate change, biodiversity, climate change impacts, observations, problems, solutions, Cameroon Highlands forests.

Inroduction

The Cameroon Highlands possess a unique flora and fauna, with numerous endemics particularly among the birds and vascular plants. The high and often very localized degree of endemism is believed to have arisen during the Pleistocene age when climatic changes forced the forest to retreat to the wetter mountainous areas of the African continent (figure 1), leaving tracts of forest or "refugia" cut off from one another. Within Cameroon, the most important refugia are thought to have existed in the areas surrounding Mount Cameroon, Mount Kupe and the Bamenda Highlands (Figure 2). The history of the distribution and evolution of montane forest species has been reviewed by Moreau (1966), Hamilton (1976, 1982) Diamond and Hamilton

(1980) and Stuart (1986). It is now generally believed that past climatic changes have been instrumental in determining the present distribution of forest species, particularly of birds (Moreau 1966; Hamilton 1976). Prior to 12,000 B.P. the climate seems to have been cold and dry. This must have resulted in the distribution of montane forest in the Cameroon Highlands much different from that seen today. Since it was much colder it is likely that the montane forest were displaced to lower altitudes. Since it was also drier these forests were probably more restricted than they are today (Livingston 1975; Hamilton 1982). It is generally believed that the wettest places today are likely to be the wettest places during this period (Livingston 1975). In West Cameroon, the wettest places are Mount Cameroon, Mount Kupe, Mount Nlonako and the Bamenda Highlands (Figure 2). It is therefore likely that montane forest were restricted to these lower mountain slopes during the last glaciation, and became important forest refugia for many montane species of plants and animals.

The restricted and isolated distribution of small populations of montane flora and fauna in forest refugia is believed to have resulted in specification and subspeciation, since the period of isolation was fairly brief (around 20,000 years before the temperatures rose and the forest expanded) most differentiation was at subspecies level. Many of the montane forest species have been unable to dispense subsequently from the refugia and colonise other mountains.

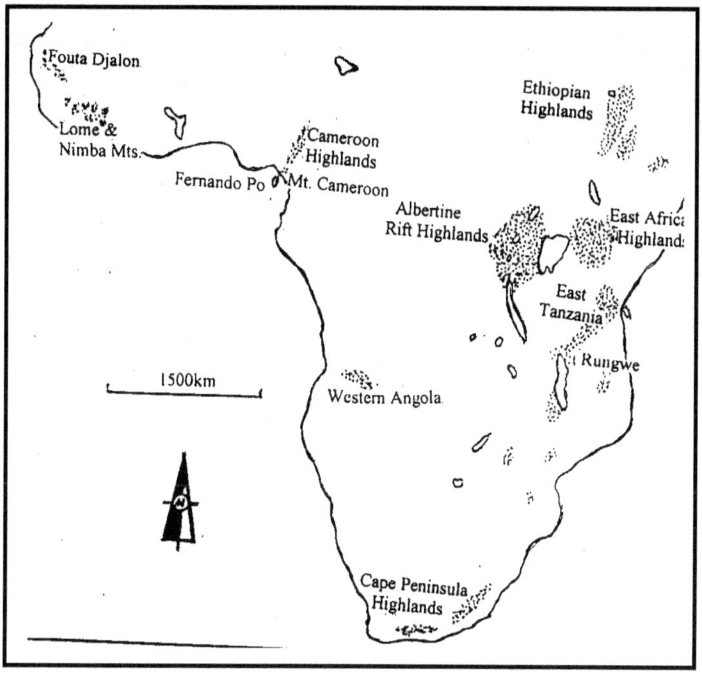

Figure 1: *The distribution of Afromontane forest refugia in Africa.*

When the climate became more mild and humid, about 12,000 years ago, montane forests were probably displaced up to higher elevations by lowland forest, while increasing humidity made it possible for the forest to expand beyond the refugia to mountains previously too dry to support forest. The expansion of the forest could have increased the opportunities of colonization for long distance wanderers. This could explain the disjunct distribution of afromontane vegetation and birds in East and West Africa. Whatever the explanation, the Cameroon Highlands exhibits the western limits of typical members of East African fauna and possess a vegetation with many East African affinities, particularly the *Podocarpus/Arundinaria* formation which is widespread in East Africa, but now found nowhere else in West Africa (Macleod 1986). The chapter examines the biodiversity features, makes observations on the current status of the forest, the types and severity of threats and establishes how climate change affects biodiversity. It finally, identifies the scope for mitigating climate change as well as supporting traditional societies in sub-Saharan Africa in adapting to it.

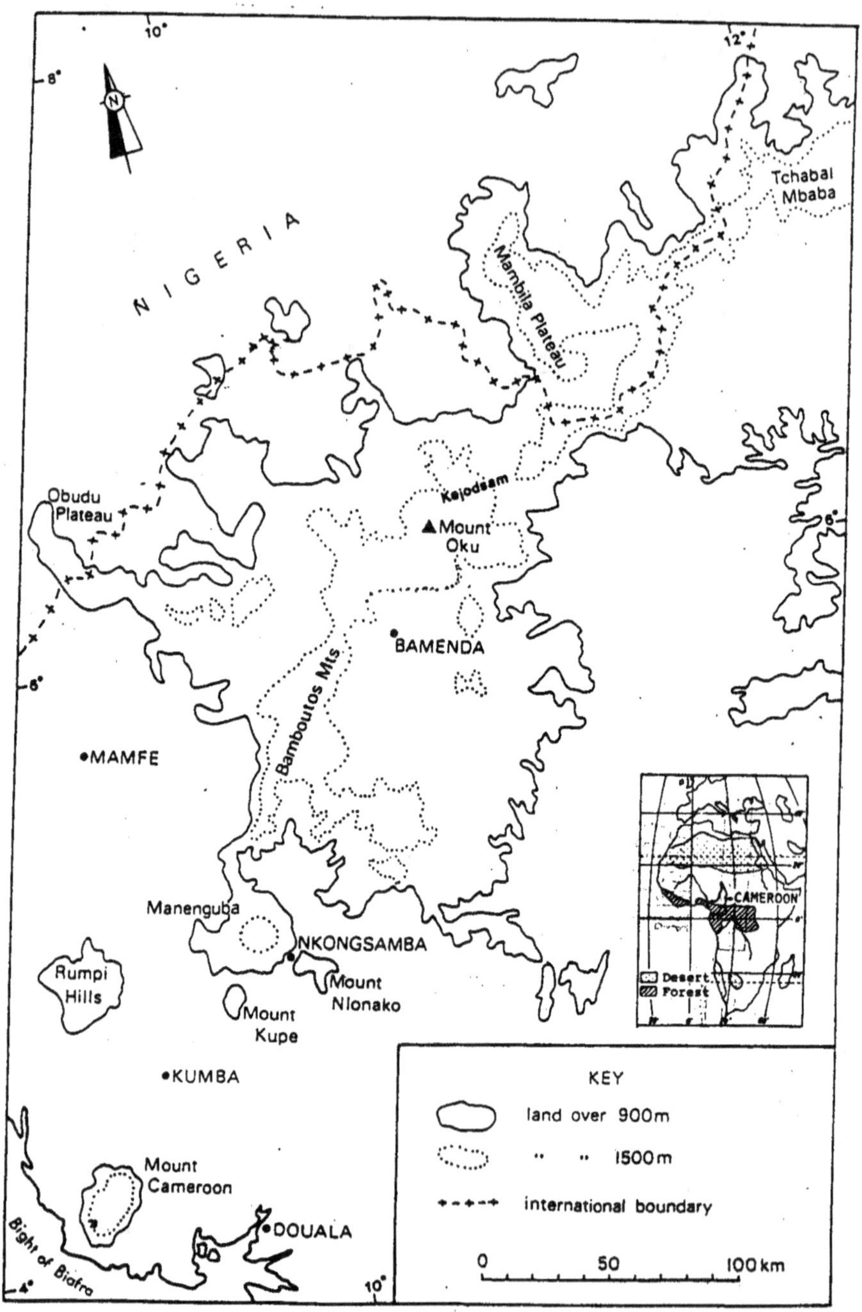

Figure 2: Location of the West Cameroon Highlands.

Geographical Setting of the Ecoregion

Geographically this ecoregion encompasses the mountains and highland areas of the border region between Nigeria and Cameroon, excluding Mount Cameroon. Most of the ecoregion lies within a narrow rectangle of some 180 km by 625 km, oriented southwest to northeast and originating about 50 km inland of Mount Cameroon. At the southern extremity, the ecoregion covers the Rumpi Hills, the Bakossi Mountains, Mount Nlonako, Mount Kupe and Mount Manengouba. It then extends northeast towards the Mambila Plateau, with an extension north to the Bamenda-Banso highlands and outliers northeast to Mambila and northwest to the Obudu Plateau of Nigeria. The main trend of the ecoregion continues north-east along the western flank of the Adamawa Plateau to Tchabal Gangdaba with two small outliers further east (Stuart 1986, Gartlan 1989, Stattersfleld *et al.* 1998). The highest point within this ecoregion is Mount Oku at 3,011 m in the Bamenda-Banso highlands. Most of the remainder is below 2,600 m in elevation. At about 800 m to 1,000 m the ecoregion grades into lowland vegetation communities of other ecoregions. In the majority of cases, however, the lower boundary of these forests is now determined by conversion to agricultural land. Although located in tropical Africa the mean maximum temperatures are below 20°C due to the effects of altitude. At the southern extremity, closer to the coast, rainfall is around 4,000 mm per annum, declining inland to 1,800 mm or less.

These mountains form part of a chain of former volcanoes that stretch inland from the sea. While there are currently no active volcanoes within the ecoregion, the effects of vulcanism are still present. Soils derived from these volcanoes are fertile, which makes the land attractive to farmers. Combined with adequate rainfall, this contributes to a high human population density. In White's (1983) phytogeographical classification, these mountain areas fall within the Afromontane archipelago-like regional centre of endemism that spans the entire continent. The elevational stratification that occurs along these mountains is quite distinct. Vegetation consists of submontane forests between 900 and 1,800 m, and above this a mixture of montane elements, including distinct montane forests and patches of montane grasslands, bamboo forests and subalpine communities. Five species of tree characterize the forested montane zone: *Nuxia congesta, Podocarpus latifolius, Prunus africana, Raparzea melanophloeos,* and *Syzygium guineense bamendae*, and these trees become increasingly covered with an epiphytic flora, especially orchids and mosses, at higher altitudes. Other important montane species include *Crassocephaluim mamnii, Hypericum lanceolatum, Myrica humilis, Piilippia mannii, Schefflera abyssinica.*

Observations on Biodiversity Features

Although plant diversity and biogeography are poorly known, tree species diversity tends to be low, but the diversity of non-woody plants such as grasses is high. Highest levels of tree endemism are found in the submontane region and the higher elevations of the adjacent Mount Cameroon and Bioko Montane Forests (Figure 2) ecoregion. Overall tree diversity, however, decreases with increasing elevation. There is also a significant endemic flora in the grasslands, heathlands, moorlands and other non-forested habitats at higher elevations. These non-forested habitats share elements with the high mountain plant communities to the east, namely the Ruwenzori-Virunga montane moorlands. The Cameroon Highlands ecoregion is not high enough to contain pure Afroalpine elements, which are restricted to the mountains of East Africa (Figure 1)

The forests have affinities with the highland forests of Angola on the southern fringes of the Congo Basin (Kingdon 1990) and especially with forests in East Africa. Most of the trees in the montane forests of Carneroon are also present in the mountains of eastern Africa, for example *Alangium chinense, Albizia gummifera, Apodytes dimidiata. Cassipourea gummiflua, Croton macrostachyus, Ilex mitis, Olea capensis, Podocarpus latifolius, Polyscias fulva, Prunus africana, Schefflera abyssinica, Strombosia scheffleri, Xymalos monospora,* and at edges *Agauria salicfolia, Maesa lanceolata, Myrica humilis, Nuxia congesta, Pittosporum viridUlorum, Rapanea melanophloeos and Scolopia zeyheri* (Dowsett-Lemaire 1989). The liane *Dregea schimperi* (Asclepiadaceae), common in the drier type of montane forest of this ecoregion is widespread in the mountains of East Africa (Thomas and Thomas 1996).

The ecoregion contains exceptional levels of avian endemism,. There are seven strictly endemic bird species, the Bamenda apalis (*Apalis bamendae*), Bangwa forest warbler (*Bradypterus bangwaensis*), whitethroated mountain-babbler (*Kupeornis gilberti*, EN), banded wattle-eye (*Platysteira laticincta*, EN), Bannerman's weaver (*Ploceus bannermani*, VU), Bannerman's turaco (*Tauraco bannermani*, EN) and Mt. Kupe bushshrike (*Telophorus kupeensis,* EN) (Bowden and Andrews 1994, Stattersfield *et al.* 1998). An additional nine montane endemics in common with Mt Carneroon and Bioko, these are *Psalidoprocne fuliginosa, Andropadus tephrolaemus, Phyllastrephus poensis, Phylloscopus herberti,*

Urolais epichlora, Poliolais lopezi, Nectarinia oritis, Nectarinia ursulae, Nesocharis shelleyi. There are also 14 species in common with just Mt Cameroon and not Bioko, adding *Andropadus montanus, Phyllastrephus poliocephalus, Laniarius atroflavus, Malaconotus gladiator, Cossypha isabellae* and the race *Cisticola chubbi discolor* (sometimes considered a separate species *C. discolor*) (Dowsett 1989, Dowsett-Lemaire and Dowsett 1989, 2000, Fotso et al. 2001).

Eleven small mammal species are considered strictly endemic to this region: Eisentraut's striped mouse (*Hybomys eisentrauti*, EN), an African wood mouse species (*Hylomyscus grandis*), Mount Oku mouse (*Lamottemys okuensis*, EN), Mittendorf's striped grass mouse (*Lemniscomys mittendorfi*, EN), two (*Lophuromys dieterleni* and *L. eisentrauti*), Oku mouse shrew (*Myosorex okuensis*, VU), Rumpi mouse shrew (*M. rumpii*, CR), western vlei rat (*Otomys occidentalis*, EN), Hartwig's soft-furred mouse (*Praomys hartwigi*, EN), and Isabella's shrew (*Sylvisorex isabellae*, VU). In addition to these smaller species, there is also an isolated population of an endemic subspecies of lowland gorilla (*Gorilla gorilla diehli*, EN). There are also populations of rare larger mammals — for example there are several groups of drill (*Mandrillus leucophaeus*, EN), in Bakossi, as well as healthy populations of Preuss's red colobus (*Procolobus pennanti preussi*, EN) and chimpanzee (*Pan troglodytes*, EN). One large group of drill was seen on Mount Kupe as recently as 2000.

Very high levels of endemism are observed among amphibians, with nearly 40 species as strict endemics (Gartshore in Stuart 1986): *Hylarana longipes, Petropedetes parkeri, P. perreti, Phrynobatrachius cricogaster, P. steindachneri, P. werneri, Phrynobatrachus sp.* (Oku), *Phrynodon Sp.* 1 *sensu* Amiet (1975), *Phrynodon sp.* 2 *sensu* Amiet, *Arthroieptis adoififriedericii, Cardioglossa melanogaster, C. oreas, C.*

Among the reptiles the following nine species are considered narrow endemics: *Atractapsis coalescens, Chamaeleo eisentrauti,* Pfeffer's chameleon *(C. pfefferi),* four-horned chameleon *(C. quadricorrnis)*, *Cnemaspis gigas, Leptosiaphos chriswildi, L. ianthinoxantha*, angel's five-toed skink (*L. lepesmei*) and *Panaspis duruarum.* In addition to the narrow endemics, there is also a significant overlap between the flora and fauna of this ecoregion and that of the nearby Mount Cameroon and Bioko Montane Forests ecoregion. Fifty near endemic plant species are shared between the two ecoregions. Similarly, 27 bird species are confined to these two ecoregions (Stattersfleld et al. 1998). A similar pattern of overlap in taxa is seen in the plants, birds, reptiles, amphibians, invertebrates and small mammals

Observations on the Current Status, Types and Threats

While the different mountains and highlands that comprise this ecoregion are naturally geographically isolated, human activities are increasingly fragmenting, degrading and isolating the remaining forest patches. The area around Mount Oku in the Bamenda-Banso Highlands supports some 100 km^2, and there are other patches in this region. The Bakossi Mountains have at least 200 km^2 of mid-altitude and montane forest above the altitude of 1000m; and the lowland forest ("Western Bakossi") covers some 400 km^2. The Mt Nlonako Faunal Reserve also contains a partial forest continuum from the montane section to the lower levels, Tchabal Mbabo, in the northern sector, has almost 50 km^2 of virtually pristine montane forest (Thomas and Thomas 1996). In Nigeria, the biggest patch in the Gotel Mountains covers 46 km^2 (peak Gangirwal) and there are other patches. Montane forest remnants also remain in gullys of the Obudu Plateau and on the highland areas further to the north such as the Mambila and Mana Plateaux and Gashaka Gumti; all are smaller than the Cameroon patches (Sayer *et al.* 1992, Stattersfield *et al.* 1998).

Forest loss has been significant on many mountains in this ecoregion. Much of the Bamenda-Banso Highlands were once covered by forest, but cover has declined by more than 50% since the 1960s through conversion for cultivation, because of the relatively fertile soils and the reliable rainfall in this area (Stuart 1986, Collar and Stuart 1988, Alpert 1993). Similarly, deforestation has also occurred on the Obudu Plateau.

This is one of the least well-protected ecoregions in Africa. No part of this ecoregion is under formal protection status in Cameroon, although local traditional rulers still exert considerable authority over land use. The main section of Bakossi (550 km^2) has been proposed as "Protection Forest", banning all logging. Kupe has been proposed as a "Strict Nature Reserve", and the boundaries of this reserve were successfully delineated with the participation of the local people in 2000-2001. The forest at Oku has some form of protection and the boundaries are well demarcated; unfortunately the demand for firewood for neighbouring towns (including Bamenda) is so high that the forest may not have a long-term future. Nlonako Mountain is a Faunal Reserve, which gives it protection from logging, even if the hunting ban is not respected. In Nigeria The Gashaka Gumti National Park contains some montane forest and some fragments remain at Obudu in the Okwangwo section of the Cross River National Park.

The natural habitats of this ecoregion are highly threatened and are being lost through conversion to agriculture, unsustainable use of timber, fires from farmland, and collection of firewood and construction materials (Collar and Stuart 1988, Gartlan 1989, Alpert 1993, Blom *et al.* in prep). Because volcanic rock produces good soils, there is considerable pressure to convert areas to farmland. However, these soils also dry easily, leading to desiccation during the dry season. Firewood collection is a major cause of degradation of forest patches. Hunting also threatens the remaining larger mammals.

Several plants and animals are used for food in traditional communities. Diversity is key to nutritionally adequate diets. Adaptation to changing conditions due to climate change is better achieved through diversification, reducing the negative impacts of extreme events, like floods, droughts, or diseases. In general, wild sources of food remain particularly important for the poor and landless and play a very significant role in times of famine, and insecurity and conflict, when food supply mechanisms are disrupted and people have limited access to food. Traditional communities are also highly dependent on ethnopharmacology. Several medicinal plants have been identified by the Limbe Botanic Garden. Biodiversity contributes not only directly to food security, but also to a number of environmental services provided to agriculture, such as pollination of food crops, water regulation, disease control and others. Intact ecosystems are important sources of local livelihood. Biodiversity also serves as a safety net for vulnerable households in time of crisis because it offers income opportunities, and sustains productive agricultural ecosystems. Development actions must focus on the functional relationships and processes within ecosystems. A key question is how local livelihood improvement and food security can be linked to biodiversity conservation and land management in the buffer zones of protected areas.

Climate Change Related Impacts on the Biodiversity

Climate change has a wide range of effects on biodiversity. As plants and animals rely on specific environmental conditions, changes in temperature and humidity regimes affect the performance and competitiveness and relationships of species. Species adapted to cold retreat to higher altitudes and latitudes while species living in warm areas may expand their ranges of distribution. This can cause problems if species expand that are regarded as pests, or if invasive species do harm to locally adapt ecosystems. The desert locust is an example, a slight increase in precipitation in susceptible areas in Africa where this insect lives may cause increased population densities (Cheke and Tratalos 2007 cited by Vohland and Lotze-Campen 2008).

However, shifts in distribution areas will not always be possible as species may not be able to migrate through a landscape increasingly fragmented by human activities (Figures 3, 4, and 5). They may not also find a new place to establish as climate regimes do not fit or land use is inadequate. Consequently, climate change must be expected to further increase the global rate of species extinction.

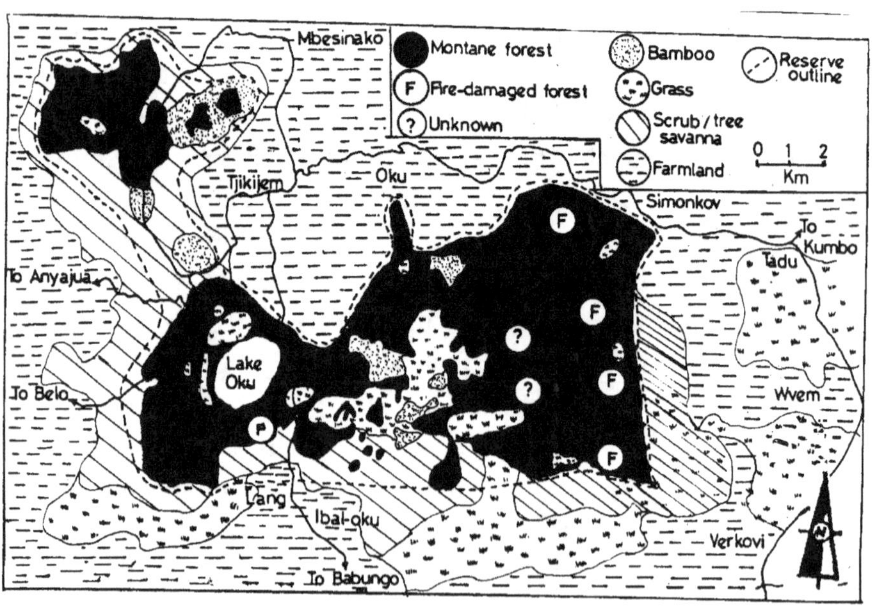

Figure 3: *Degradation of Mount Oku forest. Source: Macleod, 1986.*

The rapid change of environmental conditions is of particular concern for some hotspots of species diversity in Africa. These areas could transform from tropical forests to a Savannah-like formation. These areas will then lose large numbers of species, many of which are still unknown. These species form part of the living library needed to detect new pharmacological or crop species.

According to Vohland and Lotze-Campen (2008) on a single plant level. Climate change impacts a plant's phenology, that is, the time of the year when it starts flowering, develops fruits or dies. In Sub-Saharan Africa, the increased variability of precipitation might decrease the number of plant generations per year. In many areas, increasing temperatures are combined with a decreasing or at least more variable supply of water.

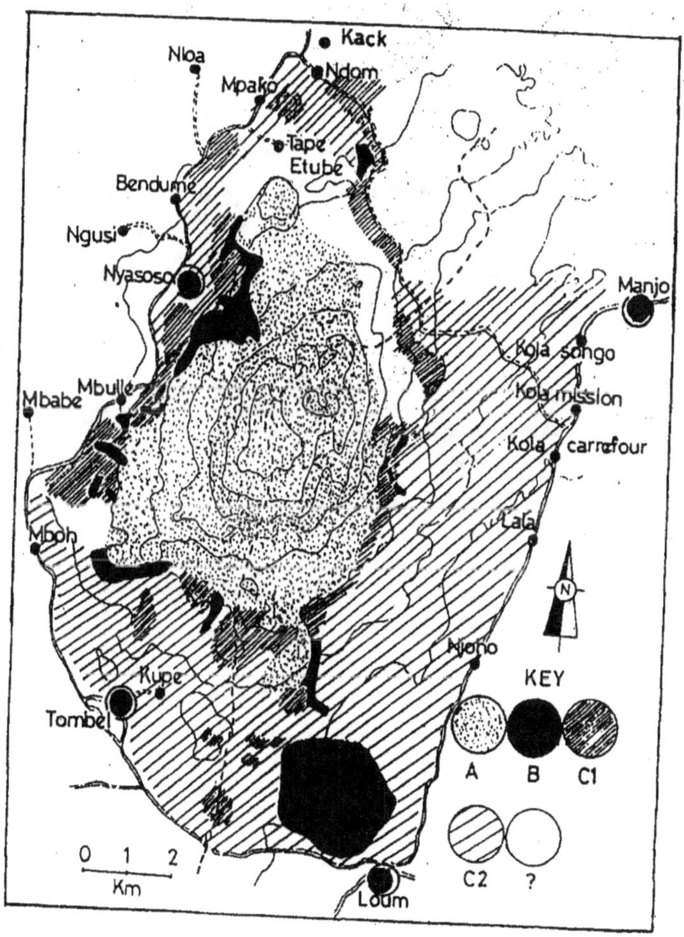

Figure 4: *Degredation of Mount Kupe Forest: A= primary forest, B= secondary or disturbed forest, C1 = Farms with a less than 50% canopy cover, C2 = Farms with a few or more trees,? = areas yet to be mapped.*
Source: *Mount Cameroon Project*

Another climate change related impact described by Vohland and Lotze-Campen (2008) is the potential "fertilization effect" of increased atmospheric

CO_2. Many plants are limited in their growth by CO_2, therefore increased supply may lead to increased biomass accumulation or enhanced phenonological development- as long as water and other nutrients are not limiting. In principle this is true for crops, too, but recent research results indicate that the delivery value of some crops might be reduced as the supplementary CO_2 leads to more sugar instead of proteins (Stafford 2007 cited by Vohland and Lotze-Campen 2008)

Decreasing biodiversity may be related to reduced food security as varying conditions cannot always be buffered by genetic or species diversity. The risk of famine increases with climate change, particularly when farming systems and livelihood options rely on few or only one species and institutions cannot provide support. Clearing mountain forest as shown in figures 3, 4 and 5 reduces the water storage functionality in catchments. In combination with floods and droughts induced by climate change, people suffer from erosion and water depletion.

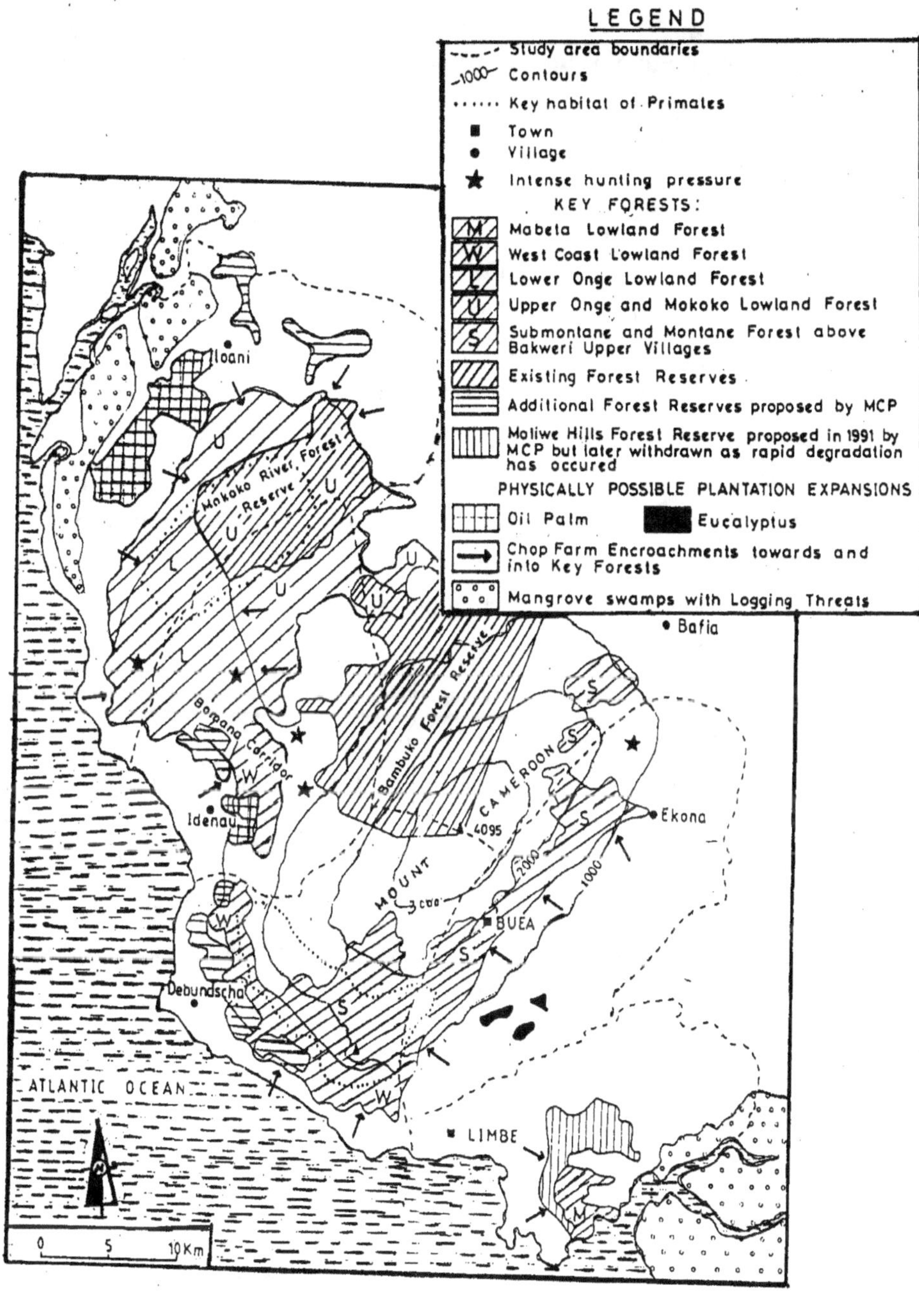

Figure 5: *Conservation values and threats in the Mount Cameroon Region. MCP = Mount Cameroon Project.*

Source: *Mount Cameroon Project.*

A likely impact of climate change on mountain environments is upslope altitudinal migration of vegetation climatic belts. Thus affected zones will experience increased vulnerability (Parry, 2000). Migration of vegetation zones around mountains to a different aspect is also possible, with implications for certain floral assemblages. Upslope migration will probably result in the contraction and fragmentation of populations of plants and associated fauna in the present mountain zones. A key factor in the survival of plant species will therefore be their dispersal capabilities (McGlone *et al.*, 2001). This tends to be low and therefore can make montane species particularly sensitive to changes in climate. Low dispersal capabilities and other characteristics such as edge of their range, geographically localized, genetically impoverished, slow producers or highly specialized are key characteristics that make them particularly sensitive to climate change (McNeely, 1990).

There are a host of other stresses that may affect montane systems as climate change progresses. These include:

- Soil changes, with influences on the growth of plants (Kundzewicz and Parry, 2001).
- Changes in fire frequencies, with effects on grasslands and forest ecotones (Gitay *et al.*, 2001).
- Changes in cloudiness, humidity and rainfall in areas covered by tropical montane cloud forests and resultant changes in these ecosystems and associated hydrological regimes (Foster, 2001).
- Increasing wind velocities, especially associated with intensified tropical cyclones (Giorgi and Hewitson, 2001) affecting evaporation rates and fires probabilities.
- Changes in populations of insects and diseases and their impacts on host plant or wildlife populations, and the effect of the spread of malaria and other diseases on tourism and travel markets in mountain areas currently free of these diseases (McMichael and Githeko, 2001).

Strategy Design

Policy actions must start by identifying what has to be protected: which species, species assemblages, habitats, or landscapes should take priority. To understand the likely vulnerabilities of natural and cultural resources to climate change, it is appropriate to conduct an integrated assessment (Yin and Cohen, 1994) or climate sensitivity analysis (Peine and Berish, 1999), including the following stages:

- Identify ecosystems, ecotones, species and ecosystem processes that are particularly sensitive to climate change.
- Evaluate protected area and forest resource monitoring programmes as to their capability to detect ecosystem dynamics associated with climate change;
- Utilize the results of the analysis for education materials and services;
- Identify the significant non-climatic related threats and plans to minimize them.

The possible adaptation options required to increase resistance and resilience of montane ecosystems include the following (WWF, 2003):

- Protecting adequate and appropriate space: Given the prospect of upslope shift of habitat space, there is need to ensure that each mountain protected area has as great a range of elevations, slope aspects, and habitat mosaics as possible. This requires adequate connectivity within the landscape.
- Conservation networks: This strategy seeks to protect landscapes, habitats and species threatened by climate change. Adaptation is allowed in such a network through species migration via buffer zones, protected corridors along upland water courses and planted forests. These should be designed to maximize connectivity in mountain areas where spatial variability and migration routes for range-shifts are limited.
- Participation and active management: The active involvement of land users within and adjacent to mountain protected areas in conservation is required. This is important where spatial networks are established. This should include the coordination of a number of approaches aimed at improving local livelihoods in order to decrease dependence on natural resources.
- Monitoring: A key element of any adaptation strategy must be to monitor both its implementation and the changes in the physical and biological environment. Monitoring should be interdisciplinary and integrated. It should include:

- Long-term monitoring and analysis of indicators of environmental change in mountain regions;
- Integrated model-based studies of environmental changes in different mountain regions;
- Process studies along altitudinal gradients and in associated headwater basins;
- Sustainable land use and natural resource management.

- Policies: The essential underpinnings to all adaptation options comprise appropriate plans and policies, both for specific areas and for sectors and agencies. These need to identify problems and priorities, and include appropriate legal provisions and economic instruments to ensure implementation.
- Eliminating non-climate change stresses: The anthropic activities imposing stresses on biodiversity should be controlled. Even human activities that have been sustainable in the past may cease to be so as climate change alters mountain ecosystems, making them less resilient than other terrestrial ecosystems even under the best of conditions.

The above strategies are necessary for a sound biodiversity governance in the face of climate change. Adequate information and appropriate plans and policies, including economic instruments are essential for adaptation. Individual protected areas require effective management plans, which are implemented, but it is critical that these are embedded in, and supported by government policies.

Implications for Rural Development

The problems presented by climate related impacts are challenges for rural development and poverty alleviation. The biggest challenge is to tackle increasing rainfall variability, that is, the problem of droughts and floods combined with the overall loss of biodiversity. These are not only caused by climate change. Key areas for adaptation to climate change include;

- The design of guidelines for bioenergy production: rural areas in Sub-Saharan Africa face a new period of agricultural intensification for a rapidly growing population and the production of biomass for energy. In order to safe forest refugia bioenergy production might support rural development and maintain biodiversity on a small scale, especially on a farmer-scale.

- Protection of forest and trees. One of the key roles of vegetation with respect to climate change is to buffer against increasing temperatures. Respiration of forest enhances the cooling of the surrounding area and may even increase the probability of precipitation. The delineation of areas for protection forests should have the following considerations.
  - Delineation of montane forest areas for protection above 900m elevations for capturing sub montane and montane forest habitat and subalphine communities.
  - Delineation of protection forest in the lower elevations in order to capture elevational migration of species as well as transitional zones that appear to be crucial component of speciation and of the development of endemism (Schneider *et al.* 1999)
- The development of agroforestry systems: This should seek to combine biodiversity conservation and climate change adaptation. Agroforestry systems are more resilient against climate change than treeless fields, as the trees reduce wind speed and therefore evaporation, and trap water and nutrients. They also harbour highest species diversity including predators that check insect pests
- Enhancement of genetic crop diversity: This involves the development of a high agro-biodiversity which allows for local adaptation processes. It will require that farmers be supported in maintaining their locally adapted native varieties as an insurance against future increases of temperature variation and the occurrence of new pest and diseases.
- Development and promotion of soil and water conservation strategies: soil tillage in combination with high temperatures leads to a rapid oxidation and loss of soil carbon. Organic farming and soil conservation practices such as mulching or zero tillage enhance both the soil biological diversity, its fauna and also reduce carbon losses from the soil.

Paying attention to the conservation of soils will permit a more sustainable agriculture to face climate change.

- Promotion of biodiversity governance at the global, regional and national levels. Biodiversity governance must be an intrinsic part of international and national legislation. Protected areas as a means to conserve the library for future crops and pharmacological development as well as retreat areas for plants and animals must be supported by international funding and trading mechanisms.

Nature conservation funds and "payments for ecosystem services" in sub-Saharan Africa should be a global task; it should not be regarded as the responsibilities of the countries concerned. Such payments should encourage private land users to introduce environmentally friendly land use practices. The aim is to ensure that sustainable land resource management and conservation apply not only to specially designated protected areas, but also on productive land. The concept is based on the innovative idea of making payments for ecosystem services. This is particularly important because sub-Saharan Africa faces additional challenges such as poverty and livelihoods sustenance:

- Poverty exacerbates competition for scarce resources and mean that natural resource conservation and biodiversity protection have relatively low priority.
- The creation of protected areas is usually associated with high opportunity costs; because these areas could often also be used for food production and other livelihood activities.

Conclusion

The Cameroon Highlands ecoregion comprises montane forest/grassland patches mainly above 900m elevation, scattered in an archipelago along the border area between Nigeria and Cameroon. Habitat ranges with increasing altitude from Sub-montane to montane and ultimately sub-alpine grasslands. The forest and grasslands contain exceptional levels of endemism in all taxa. In most areas these are poorly protected. Due to population pressure the remaining habitats, especially the forest, are threatened by agricultural expansion. Climate change also has a major impact on the biodiversity. Sustainable biodiversity management through the measures outlined above can play an important role in mitigating climatic change, buffering the influence of climate change as well as supporting rural communities in Sub-Saharan Africa in adapting to it.

Acknowledgments

The authors acknowledge the use of descriptions of the Cameroonians Highlands forests (ATO103) of WildWorld: www.Worldwildlife .org

References

Alpert. P. 1993 Conversing biodiversity in Cameroon. Ambio 22:44-49

Amiet, J-L. 1975 Ecologie et distribution des Amphibiens Anoures de la region de Nkongsamba (Cameroun). Ann. Fac. Sci. Yaounde 20:33-107.

Amiet, J,-L. and F. Dowsett-Lemaire. 2000. Un nouveau Leptodactylodon de la Dorsale camerounaise (Amphibia, Anura). Alytes 18:1-14.

Bowden, C.G.R., and S.M. Andrews. 1994. Mount Kupe and its birds. Bulletin African Bird Club 1:13:16.

Collar, N.J., and S.N Stuart. 1988. Key forests for threatened birds in Africa. ICBP, Cambridge, UK

Diamond, A.W and Hamilton, A.C 1980. the distribution of forest passerine birds and Quaternary climate change in tropical Africa. Journ. Zool. London p 379-402

Dowsett, R.j., (ed). 1989. A preliminary natural history survey of Mambilla Plateau and some lowland forest of eastern Nigeria. Tauraco Research Report No. 1, 56 pp. Tauraco Press.

Dowsett-Lemaire, F. 1989, The Flora and phytogeography of the evergreen forests of Malaw. 1: Afromontane and mid-altitude forests. Bull. Jard. Bot. Nat Belg. 59:3-131.

Dowsett-Lemaire, F. and R.J. Dowsette, 1998. Zooological surveys of small mammals, birds and frogs in the Bakossi and Kupe Mts, Cameroon. Unpublished report for WWF-Cameroon, 46 pp.

Dowsett-Lemaire, F. and R.J. Dowsette, 2000. Further biological surveys of Manenguba and Central Bakossi in March 2000, and an evaluation of the conservation importance of Manenguba, Bakossi, Kupe and Nlonko Mts, with special reference to birds. Unpublished report for WWF-Cameroon, 45 pp.

Fotso, R.,F. Dowsett-Lemaire, F. and R.J. Dowsette, Cameroon Ornithological Club, P. Scholte, M. Languy, and C. Bowden. 2001. Cameroon. Pp 133-159 in L.D.C Fishpool and M.I. Evans (eds). Important birds areas of Africa and associated islands: priority sites for conservation. Bird Life Conservation Series No. 11 Pisces Publication and BirdLife International, Newburg and Cambridge, UK.

Gatland, S. 1989. La conservation des ecosystems forestiers du Cameroun. IUCN, Gland and Cambridge.

Giorgi, F. and Hewitson, B (2001) Regional Climate Information – Evaluation and projections. In : J. T. Houghton, Y. Ding, D. J. Griggs, M. Nougier etc. (eds). *Climate change 2001: the Scientific basis.* Contribution of Working Group I, 3rd Assessment Report of IPCC, Cambridge University Press, p. 583 – 638

Gitay, H., Brown, S., Easterling, W. and Jallow, B. (2001) Ecosystems andtheir goods and services. In: O. Canziani, O. F. Dokken, D. J. Leary, W. A. McCarthy, and K. S. White (eds). *Climate change 2001: Impacts, adaptation and vulnerability.* Contribution of Working Group II, 3rd Assessment Report of IPCC, Cambridge University Press, p. 235 – 342

Hamilton, A.C. 1976. The significance of patterns of distribution shown by forest plants and animals in tropical Africa for the reconstruction of Upper Pleistorene Palaeoenvironment: a review. Paleoecol. Afr. Vol 9, 83-97

Hamilton, A.C. 1982, Environmental history of East Africa: a study of the Quaternary. Academic Press, London

Kundzewiez, A. and Parry, M. (2001) Europe. In: O. Cunziani, Dokken, D. J. Leary, N. McCarthy, and K. S. White (eds). *Climate change 2001: impacts, adaptation and vulnerability.* Contribution of Working Group II to the 3rd Assessment Report of the IPCC, Cambridge University Press, p. 641 – 692

Kingdon, J. 1990. Island Africa; the Evolution of Africa's Rare Animals and Plants. Collins, London

Livingston, D. A. 1975. Late Quaternary climate change in Africa. Ann. Rev. Ecol. Syst. Vol.6. p 249-248

Macleod, H. 1986, The conservation of Oku Mountain forest. ICBP, Cambridge, p 33-35

McGlone, M.; Duncan, R.; and Heenan, P. (2001) Endemism, species selection and the origin and distribution of vascular plant flora of *New Zealand. Journal of Biogeography, Vol. 28*, p. 199 – 216

McMichael, A. and Githeko, A. (2001) Human health. In: O. Cunziani, Dokken, D.J. Leary, N. McCarthy and K. S. White (eds). *Climate change 2001: contribution of Working Group II to the 3rd Assessment Report of the IPCC,* Cambridge University Press, p. 451 – 485

McNeely, J. A. (1990) Climate change and biological diversity. In: M. Boer and De Groot E. S. (eds). *Landscape ecological impact of climate change.* IOS Press, Amsterdam, p. 406 – 429

Moreau, R.E.1966. The bird fauna of Africa and its islands. Academic Press, London.

Parry, M. E. (2000) Assessment of the potential effects and adaptation of climate change in Europe. The Europe ACACIA Project, Jackson Environmental Institute, University of East Anglia, Norwich.

Peine, J. and Berish, C. (1999) Climate change: effects in Southern Appalachians. In: J. Pein (ed). *Ecosystem management for Sustainability: Principles and practices.* Lewis, New York, p. 397 – 415

Sayer, J.A., C.S Harcourt, and N.M. Collins. 1992. The Conservation Atlas of Tropical Forests: Africa. IUCN and Simon & Schuster,

Schneider, C.J., Smith, B. Larison, B., and C. MORITZ. 1990. A test of alternative models of diversification in tropical rainforests: ecological gradients vs rainforest refugia. Proceeding of the National Academy of Sciences, Vol.96, No. 24, Pp 13869-73.

Stattersfield, A.J., Crosby, Long, and D.C. Wege. 1998. Endomic bird areas of the world: priorities for biodiversity conservation. BirdLife Conservation Series No. 7., Bird Life International, Cambridge, UK, International Council for Bird Preservation.

Stuart, S.N. (ed.) 1986. Conservation of Cameroon montane forests. Cambridge, UK. International Council for Bird Preservation.

Thomas, D. and J. Thomas. 1996. Tchabal Mbabo Botanical Survey. Consultants' Report to WWF Cameroon Program Office. 44 pages plus

Vohland, K. and Lotze-Campen, H.2008. Biodiversity and climate change. The international Journal for Rural development, DLG-Verlays GmbH, Frankfurt.

White, F. 1983. The vegetation of Africa, a descriptive memoir to accompany the UNESCO/AETFAT/UNSO Vegetation Map of Africa (3 Plates, Northwestern Africa, Northeastern Africa, and Southern Africa (1:5,000,000). UNESCO, Paris. Prepared by: Allard Blom Reviewed by: in progress.

WWF (2003) User's manual for building resistance and resilience to climate change in natural systems. WWF Publication, New York, 240p.

Yin, Y. and Cohen, S. (1994) Identifying regional goals and policy concerns

associated with global climate change. *Global Environmental Change, Vol. 4,* p. 246 – 260

Chapter Six

Climate Change and Management of Coastal Zone and Marine Ecosystem

Summary

Tropical marine ecosystems are rich in biodiversity and provide a valuable source of protein that supports local economies. They include mangroves, sea grass, vast pelagic systems and coral reefs. Despite the goods and ecosystem services they provide, the coastal zones and marine ecosystems are under human-related stresses, climate change-related stresses and anthropogenic climate change threats. The chapter reviews the marine ecological circumstances at the Gulf of Guinea, available pelagic stocks, fishery potentials of Cameroon and the anthropogenic stresses on the coastal zone and marine ecosystems. It further assesses the anticipated effects of climate change impacts on the systems and makes an overview of the appropriate types of strategies necessary for building resistance and resilience. Finally, the chapter concludes that in addition to designing strategies for building resistance and resilience, it will also be necessary to limit the rate and extent of anthropogenic climate change.

Key Words: Climate change, coastal zone, marine ecosystem, resilience, resistance, anthropogenic stress, climate change related stresses.

Introduction

Cameroon possesses a portion of the Atlantic Ocean. The country is situated at the Gulf of Guinea and has a narrow coastal plain which hardly exceeds 150km in width. The Gulf of Guinea is rich in biodiversity and provides productivity to support human populations. They include mangrove ecosystems and pelagic systems. This marine ecosystem supports local subsistence fishing (craft fishing) as well as international commercial fisheries. The mangrove swamps in the estuarine creek systems act as a filtration system for the estuarine and freshwater. Mangroves offer protection from coastal erosion and trap sediments from terrestrial runoff. These habitats also serve as nurseries for many invertebrates and fish. The complex ecology of mangroves is well appreciated but not well understood. Efforts towards a greater understanding are constantly being challenged as these systems are altered by local anthropogenic stresses and global climate change. The resistance (ability to withstand change) and resilience (ability to recover from change) of an ecosystem determine how well it can deal with stresses. Proper management can enhance resistance and resilience to climate change by alleviating the overall pressures on the system, giving it more flexibility to mobilize its natural defences (Noss, 2001).

This chapter reviews the marine ecological circumstances at the Gulf of Guinea, the coastal and marine ecological potentials of the area and Cameroon in particular and the human-related stresses on marine ecosystems. It further assesses the effects of climate change on the systems and makes an overview of the types of strategies that might be adopted to increase their resistance and resilience to climate change.

Regional Setting of the Gulf of Guinea and the Study Area

The Gulf of Guinea is unique since it displays a zonal coastline just 5^0 off the equator and this strongly influences the distribution of currents. The distribution of water masses is further modified by large amounts of fresh water from rivers and rainfall while the only sources of high salinity are the Equatorial Undercurrent and the weak South Equatorial Countercurrent. The latter flows across the Atlantic Ocean at about 4^0S, bends southward between 10^0W and 15^0W, and when it finally reaches the African coast at about 15^0S its salinity is well below 36% and does not differ much from the waters above (Mazeika, 1968). The Equatorial Undercurrent flows eastward until it encounters Sao Tomé Island where its main part bends southward and reaches the shelf near 2^0S off Cape Lopez (Gabon). Its salinity there is still in excess of 36.2% (Rinkel *et al.*, 1966) in contrast to the surface waters which have salinities below 33% off Sao Tomé and well below 30% further inshore. The extent of the areas of low salinities is shown in Figure 1. Off the Congo, salinities of less than 20% extend in a tongue-shaped plume some 50-120 km into the ocean, depending on river outflow. Minimum salinities in the inner Gulf of Guinea vary between less than 25% in winter and 28% in summer. Little is known about the reduction of salinity off Liberia and Ivory Coast but it is likely to be about the same level (Berrit, 1969).

Water from the Gulf of Guinea can be traced along the coast by its low salinity and has been shown to reach as far as 17^{0}N, i.e. 200 km north of Cape Verde, during October/November. Its vertical extent is only some 30-50 m in the Gulf of Guinea and a strong halocline is maintained to separate it from the salinity maximum of the Equatorial Undercurrent of 40-70 m (Rinkel *et al.*, 1966; Kolesnikov *et al.*, 1971). Recently, it has been observed that under certain circumstances the Undercurrent surfaces during summer and brings cold, saline waters to the equatorial surface region. This phenomenon is not yet well understood but it might be a regular feature. Another feature is the occurrence of upwelling along the coasts of Ghana and Ivory Coast during the summer months when surface temperatures are lower than during winter. Winds in the area tend to be favourable for upwelling throughout the year, and it is not yet clear why upwelling is restricted to a well defined summer season. One suggestion is that the upwell waters are replaced by warm water adverted into the area by a regular change in the large-scale circulation.

The relatively low-lying region, interrupted by Mount Cameroon, (4,070 metres) is crossed by many swiftly flowing rivers. It stretches form Rio-del-Rey through Victoria, Tiko, Douala and Kribi down to Equatorial Guinea. The region varies in width and is marked by estuaries, creeks, swamps and sand-bars at the estuary of the Wouri River (the approach to the port of Douala). Deltas are unknown on the Cameroon coast except at the mouth of the River Mungo. The reason for this is that currents sweep away material dropped by rivers along the shore. The Cameroon coast may be divided into two: (a) the rocky coast which occurs where Mount Cameroon meets the sea and stretches from Idenau to Bimbia. It is characterized by bays (e.g. Ambas Bay), capes (e.g. Cape Nactigal), rocky islands and cliffs; (b) the low-lying coast which occurs in two different places (i) between Idenau and the estuary of the River Akpa Yafe in the north, and (ii) between Bimbia and Kribi to the south. It is characterized by mangrove swamps, innumerable creeks, sand-bars, sand pits (two north of Sanje and one north of the estuary of the River Sanaga), low-lying islands (the less marshy ones used for settlement by fishermen), and lagoons, such as that lying between Betika and Bamusso. The low-lying coastal zones are characterized by swampy or sandy nature, innumerable creeks, sandbanks, channels and low-lying islands, sand spits, lagoons and extensive mangrove forests. The coast is generally characterized by shallow waters. Estuaries and lagoons form at the mouth of the Wouri and Sanaga rivers. Other rivers in the Gulf of Guinea present estuaries and lagoons.

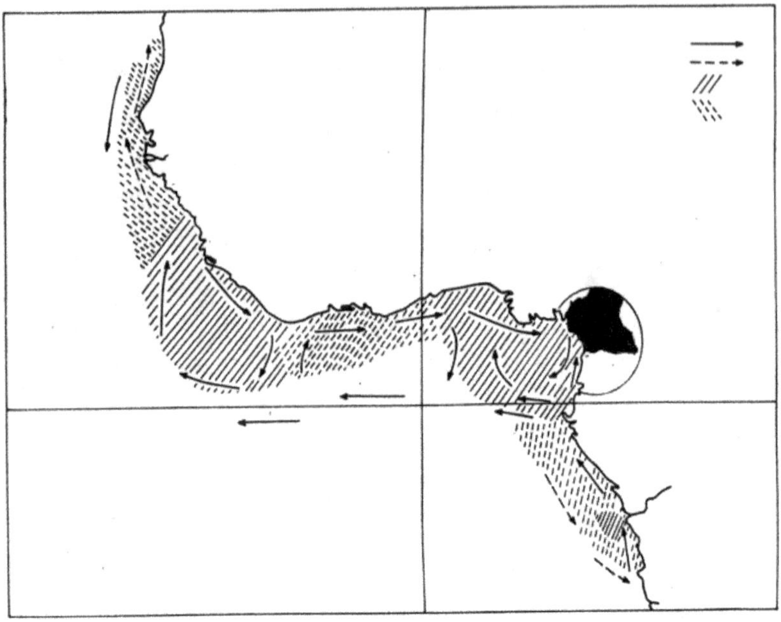

Figure 1: *Variability of currents and salinity at the surface in the Gulf of Guinea region: Broken arrows and lines designate seasonal currents and seasonally lowered salinities, respectively. The boundary of the areas of permanently low salinities roughly corresponds to the 35% isohaline during July-September, but a tongue of salinities less than 35% stretches westward from the western low salinity area for some 700 km (compare Pertroth, 1969) (from Berrit, 1969, with kind permission of Unesco Publications Office, Paris).*

Estuaries and lagoons form an important part of the coastal waters, since economic activity of the countries mostly concentrates on them. They are semi-enclosed coastal bodies of water with a temporary or permanent connection to the open sea. In these areas temperature, salinity or both are strongly influenced by local conditions, and the circulation is controlled by the resulting density distribution.

A coastal body of water which has a permanent connection with the open sea and within which sea water is measurably diluted with fresh water is called an estuary. This definition deliberately excludes areas like mouths of rivers which are sealed form the open ocean during the dry season. Bodies of water where salinity is measurably increased by excessive evaporation are also excluded and usually referred to as lagoons. This situation is somewhat regrettable since both estuaries and lagoons have in common the fact that their circulation is driven by density differences and modified by contact with the open sea (if any). In a certain sense, lagoons are inverse estuaries. Furthermore, there are a number of estuaries especially in tropical countries which change from estuarine to lagoonal characteristics with the seasons. However, under the above definition it is possible to identify several existing types of estuaries.

The "salt-wedge" estuary is found where river run-off enters an otherwise nearly motionless area. The fresh water spreads out on top of the sea water in an increasingly thinner layer as it moves seaward, surrounded by a sharp front. As the river run-sharper; as the river flow increases, it moves seaward. If the river flow is strong, the circulation changes to the "salt-wedge with entrainment" estuary where waves are set up along the interface and break upwards, mixing sea water into the top layer. Thus, salinity and transport increase seawards in the upper layer and a compensatory upstream flow develops in the salt-wedge below. The rate of increase of the downstream transport is, however, quite small. Entrainment is a one-way process by which water from the lower layer is carried into the upper layer but no fresh water is transported downward (Figure 2).

In the "partially-mixed" estuary turbulent transport of water in both directions affects the salinities of both layers and creates a continuous increase in salinity from surface to bottom and from the inner estuary toward the open sea. This type of estuary is found where the tidal flow in the estuary dominates the circulation, that is, the volume flow in the tidal oscillation is much greater than volume flow of the river. In the partially-mixed estuary the seaward flow in the upper layer increases one order of magnitude or more from the point of freshwater inflow to the outlet into the sea. A corresponding strong compensatory inflow is found in the bottom layer which can act as a sediment trap, building up shoals where the tidal oscillation is weak.

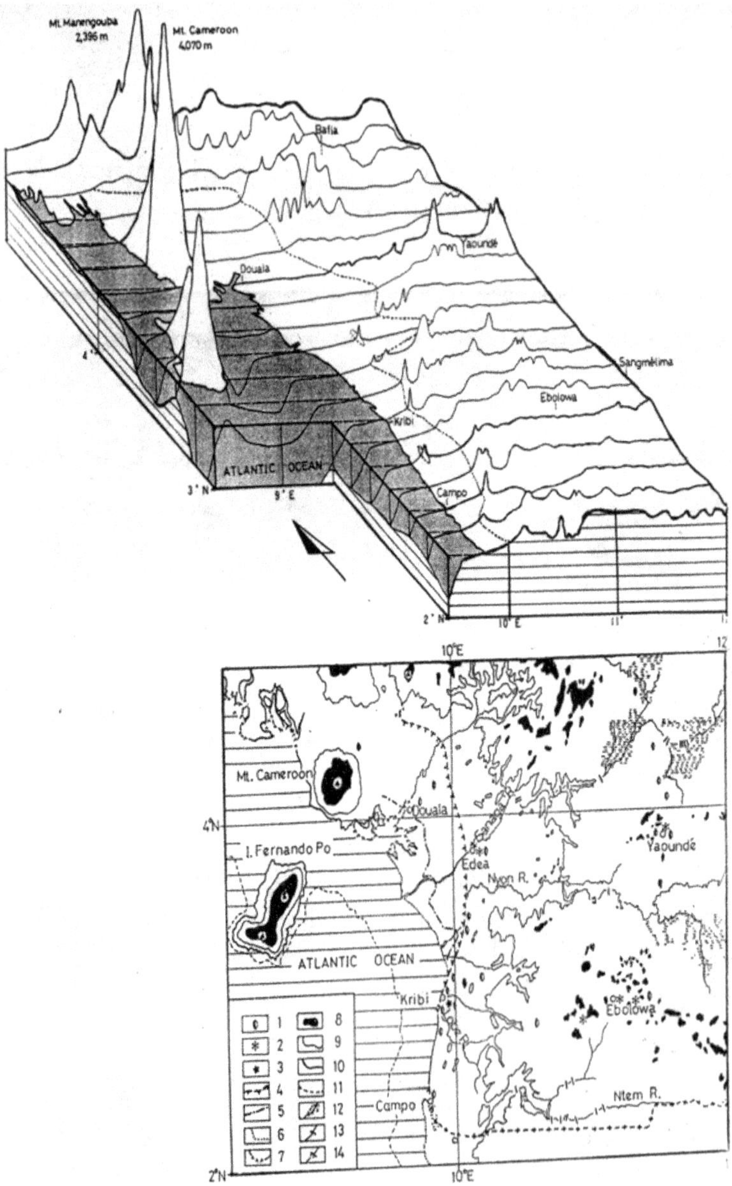

Figure 2a: *General geomorphology and geology of the coastal region of Cameroon. 1: Locality of implements collected dating back to the Bantu migrations*

into the forest, 2: Indurated cuirasses originating from weathering profile of bed rocks, 3: Secondary cuirasses originating from indurations of various kinds of gravels and sands, 4: Geological boundary between basement rocks and marine sediments since the Cretaceous, 5: Geological boundary between the Tertiary sediments and the Quaternary sediments, 6: Sandy coastline, 7: Mangrove coast, 8: Land over 900 m high, 9: Contour-line of 300 m high, 10: Present shoreline, 11: Iso-depth line of 100 m below the present sea level, 12: Swampy area, 13: Rapids, 14: Falls.

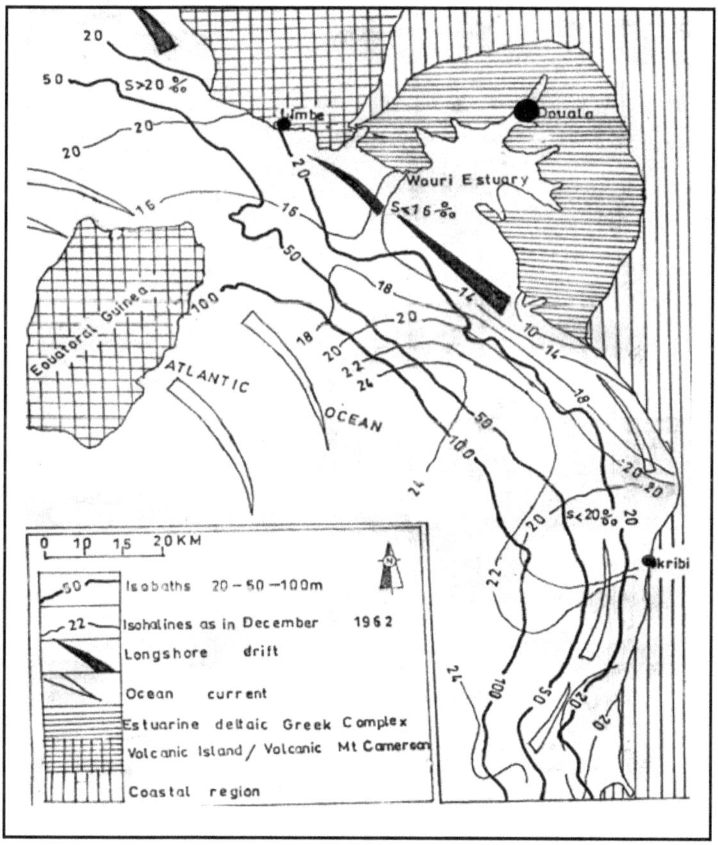

Figure 2b: *The distribution of salinity, sea depth, ocean currents and the physical setting of the coast of Cameroon (after Crosnier, 1964 cited by Morin, 1982).*

Photo: *Foreground = sandbanks, creeks and mangrove swamp forest in the estuary of River Wouri. Background = Douala city and Bonaberi Bridge across the estuary (J. A. Ngwa, 1979).*

Regional Pelagic Stocks

Bonga *(Ethmalosa fimbriata):* In West Africa outside the upwelling areas, off Sierra Leone, Liberia, Nigeria, Cameroon and Gabon, bonga *(Ethmalosa fimbriata)* is the main pelagic resource. This species is confined to the shallow brackish waters as shown in Figure 3. *Ethmalosa fimbriata* has seasonal migrations offshore and inshore in the river mouths with respect to the floods. This species is more vulnerable to pollution damage than the other pelagic fish species.

Flat sardinella *(Sardinella eba):* Flat sardinella *(Sardinella eba)* occurs in waters of low salinity but is more abundant in areas of seasonal upwelling. The large concentrations are found within the 30 m depth zone off Senegal, Guinea and Sierra Leone in the North, off Gabon, Congo (Democratic Republic of) and Angola in the South. The stock off Ivory Coast, Ghana, Togo and Benin is probably smaller. FAO (1971) reports that larvae and juveniles settling in the littoral waters off Congo (Democratic Republic of) originate from eggs laid in the area off northern Angola. Then when moving to deeper but still coastal waters, these fish are caught by the local fishery and are exploited until two years old. The fact that the older fish disappear from the Congolese fishery but are regularly caught in large quantities off northern Angola, suggests that they migrate at this age back to Angola. Similar movements by the same species are observed between Mauritania and Sierra Leone. The distribution of *Sardinella eba* along the shelf is shown in Figure 3, whereas its migration along the West African coast is given in Figure 4.

Round sardinella *(Sardinella aurita)*

The distribution of Round sardinella *(Sardinella aurita)* is in more saline waters than that of *S. eba* and the species has its major concentrations in the areas of seasonal upwelling. It is believed that the large concentrations usually remain above the continental shelf, although scattered layers have been successfully fished by mid-water trawls above depths of 1,500 and even 2,500 m. This species performs migrations parallel to the coast in relation to the seasonal movement of the boundary between the warm and the cold water in the areas of seasonal upwelling. This species is exploited near the continental edge at the end of the dry season and at the beginning of the onset of the rainy season. It then comes closer to the coast for spawning and disappears from this area in June. The stock exploited off Ghana and in the eastern waters off Ivory Coast has a more limited distribution. Off the coast of Congo (Democratic Republic of), the older fish are caught only during the upwelling season. These specimens come from and return to Angolan waters where large fish are abundant throughout the year. The distribution of *S. aurita* in respect to the shelf is shown in Figure 3 and its migration along the coast of West Africa is shown in Figure 4.

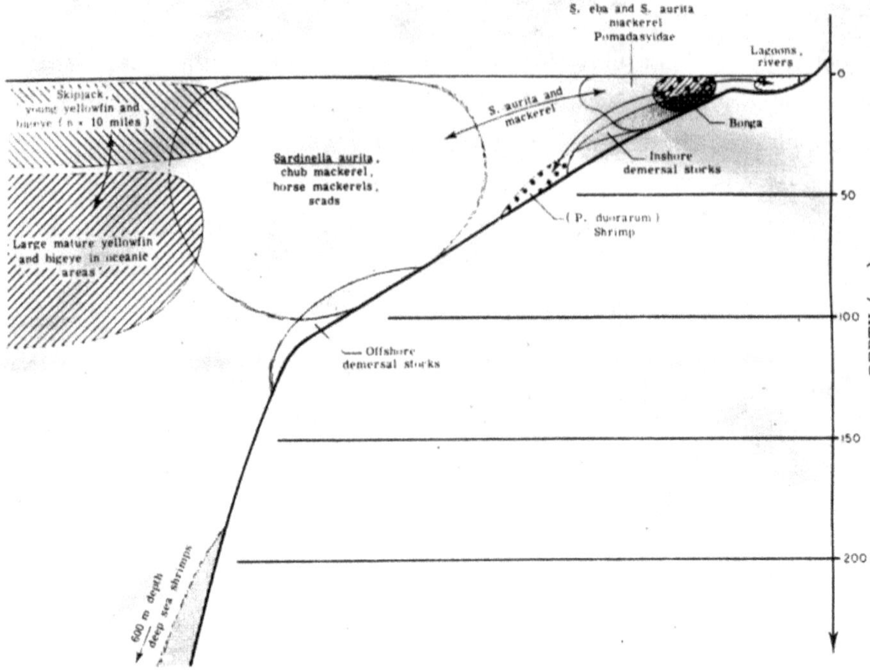

Figure 3: *The distribution by depth of fish species and species groups off the west coast of Africa between 20⁰N and 15⁰S (after Ssentongo, 1979).*

Carangids *(Carangidae)*

It is likely that chub markerel *(Scomber spp.)*, horse mackerel *(Trachurus spp.)*, scads *(Caranx, Decapterus, Selar)* and bluefish *(Pomatomus saltatrix)* have migrations along the coast, similar to those of *Sardinella aurita*, in the northern area of seasonal upwelling. Although horse mackerels and scads come closer to the coast in wet season for spawning, they do not seem to penetrate into coastal waters as do *Sardinella aurita* and chub mackerel. During the upwelling, the two last species are exploited in the inshore waters off Ghana and Ivory Coast by the local purse-seiners. Horse mackerels and scads are more commonly exploited in the vicinity of the continental edge.

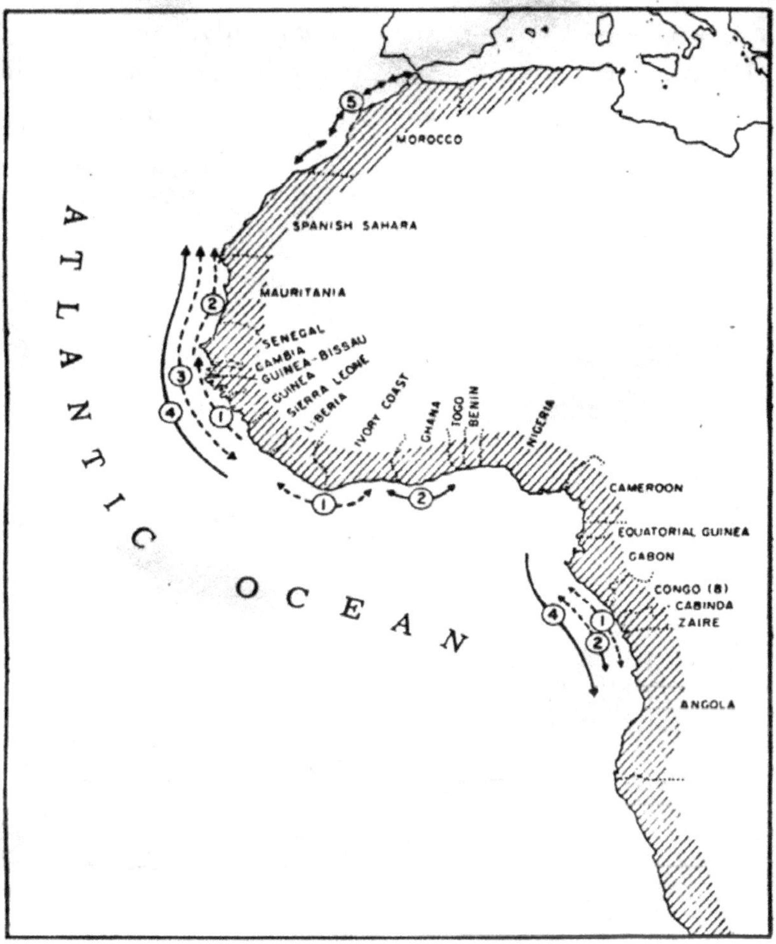

Figure 4: *Some pelagic fish species migration along the West Coast of Africa: 1 = Flat sardinella (Sardinella alba), 2 = Chub mackerels, horse mackerels, scads, bluefish, 3 = Round sardinella (Sardinella aurita), 4 = Yellowin, surface shaols, 5 = Morocco sardine (after Ssentongo, 1979).*

Marine Fish Stocks of Cameroon

Marine life in the Douala Bay and off the coast is not very diversified. There are only about 140 species. However, this number includes a large number of species of economic interest. There are mammals including some rare *Cetacean* belonging to the group of humpback whales. The rare sea cow can be found in the estuaries. Among the reptiles are five sea turtles. Sharks represent the order of *Selachians* and above all, rays such as the guitar ray and the whip-tailed stingray. Among the *stenohaline* fish are numerous types of mackerel, bonito and needlefish. The euryhaline species comprise numerous families of economic interest because of their sizes and economic value. First of all captain fish (*Polynemus quadrifilis*), barracuda (*Sphyraena piscatorum*) and the pink carp of the *lutjanus* family abounds. The *clupeidae* family comprises mainly the alosa (*Ethmalosa fimbriata* and *Ilisha Africana*). There are the species often used in the smoked fish industry. *Mugilidae* are numerous, as well as catfish of the *Arius* family. The *Scianidae* family is well known through the *corvina nigrita* (generally known as hunch back) and the *Pseudotolithus senegalensis* (bar). The *Carangidae* family contains a fairly large number of species including *Caranx carangus*, *Caranx hippos* and *Trachynotus falcatus*. The *pleuronectidae* are well represented. It must be emphasized here that in terms of species diversity the marine and brackish water population is very rich. Unfortunately, the populations of individual species are very low. This has implications for sustainable exploitation and development.

Interactions between Coastal Fisheries and Other Sectors

Figure 5 presents a model of the main interactions between coastal fisheries and other sectors. With the advent of industrialization, increasing human population and urbanization along the coast, the range of requirements for water has increased together with greater demands for high quality. It is clear that man's use of the coastal region is producing effects, which are unacceptable and may possibly lead to such ecological instability that his very survival as a species is threatened. From the foregoing review aquatic bodies support a wide array of plants, phytoplankton, zooplankton and nektonic aquatic invertebrates. The species composition and density are influenced not only by the geographical location but also by the water quality, which can be adversely, affected by human activities resulting in pollution. Aquatic environments are productive and life sustaining in terms of food resources for man.

They are also sensitive to disturbances and degradation through man's input of industrial wastes. The resulting stress causes both economic and biological losses in aquatic ecosystems (Table 1). There are several anthropic activities, which have undesirable and devastating effects on the aquatic environment. These include uncontrolled land use for urbanization or deforestation, accidental release of chemical substances, discharge of untreated wastes or leaching of noxious liquids from solid waste deposits. Similarly, uncontrolled and excessive use of fertilizers and pesticides has long-term effects on surface water resources. Ekane and Oben (2001) verified biochemical indicators along the Limbe estuary and the Douala lagoon to determine the level of pollution along the Cameroon coast. Ekane and Oben (2001) have reviewed the constraints to the development of the fishery sector in Cameroon.

Table 1: Main activities identified in the Cameroon coastal zone and their impacts on the environment.

Type of activity	Pollutants	Impact on the environment
Agro-industry Food processing	Nitrates, nitrites, Chlorine, phosphates, organic substances	Eutrophication. Decrease in yield of the food chain
Petroleum refinery Petroleum exploitation	Hydrocarbons, lead.	Accumulation of lead in the food chain. Increase death rate of fauna and flora Decrease in fish catch.
Textile Cement production Paint production Power generation Paper pulp production Soap production Bauxite processing	Acids, mercury, lead, Copper, phosphates, Asbestos, organic substances	Accumulation of lead and mercury in the food web, thereby decreasing its yield.
Battery production Engine oil production	Lead	Accumulation of lead in the food chain
Port works	Sediments	Increase in water turgidity and sedimentation increase in BODs
Production of plastics	Basic monomers	Accumulation of non-biodegradable material
Reclaiming of wetlands	Solid wastes and sediments	Loss of habitat
Domestic activities	Human wastes and	Eutrophication and

	organic substances	decrease in exygen supply
Building construction works	Solid wastes and sediments	Soil erosion,
Fishing		Loss of vegetation and habitat. Destruction of mangrove and fish nurseries

Source: Ibe et al.; 1999

Various human activities along the coast of Cameroon degrade the environment and threaten to deplete its resources. The direct threats to the environment and its resources include over-exploitation of fish and other living resources (Figure 6) and pollution from industries. Other threats include loss of vegetation cover as a result of coastal urbanization, development of touristic facilities and aqua-cultural development, which lead to erosion and siltation. There is therefore an urgent need to address the environmental issues, identify their causes and develop a strategy for their solution. The absence of coastal management and planning has led to increased pollution, erosion and destruction of living resources. With increasing coastal population pressure, this is likely to have significant long term effects on the health and quality of life of the stakeholders. Table 1 shows the main effects of various activities on the coastal environment (*Ibe et al.*, 1999).

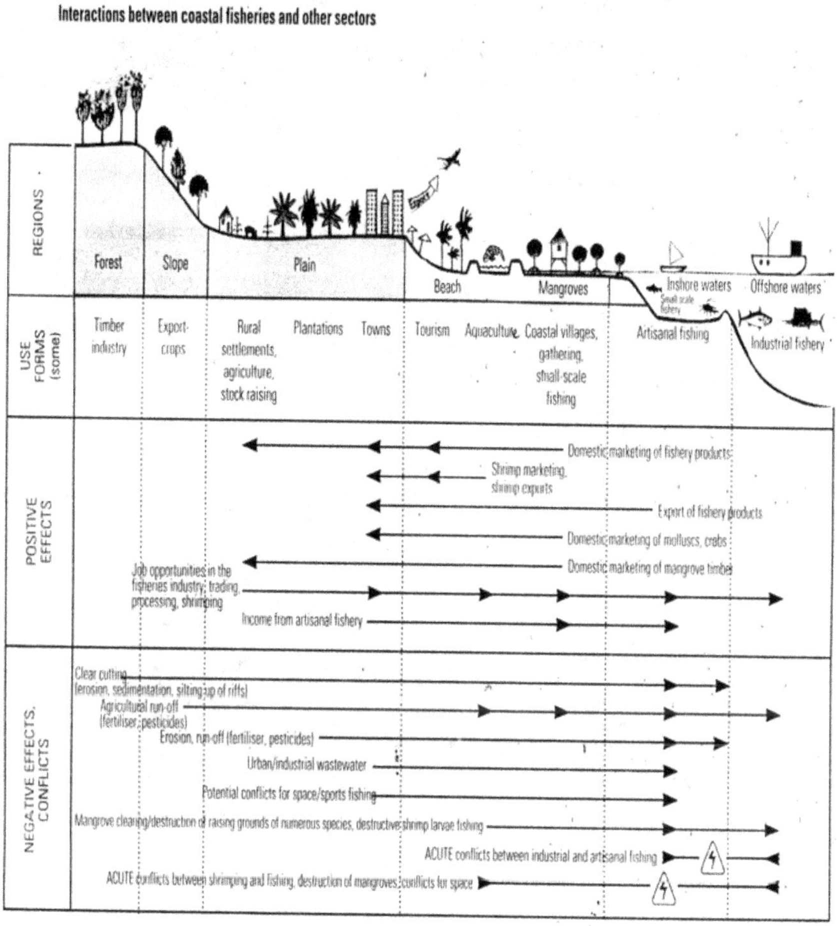

Figure 5: *Interactions between coastal fisheries and other sectors of the coastal economy.*

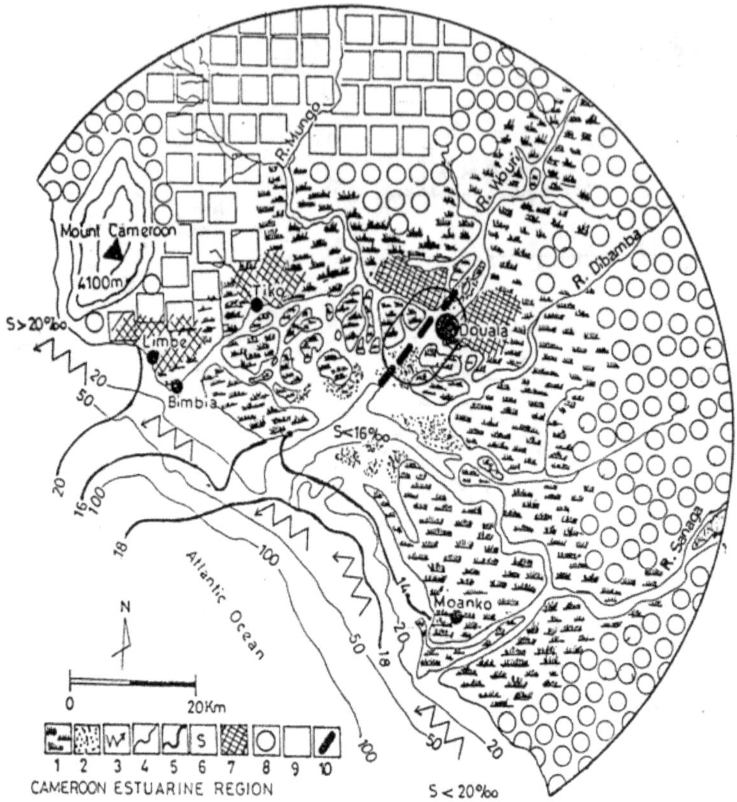

Figure 6: *Anthropic stress on Mangrove ecosystem: 1 = Mangrove swamps, 2 = Sandbanks, 3 = Longshore drift, 4 = isobath, 5 = isohaline (%), 6 = salinity, 7 = urban sprawl, 8 = slash-and-burn cultivation, 9 = agro-industrial plantations, 10 = on-going dredging and land reclamation.*

Improved planning of human activities in the coastal zone and better utilization of resources are necessary in order to address the issues of inland water, coastal and marine resource degradation. A more integrated approach to coastal area use is needed to replace current policy and institutional arrangements. Actually, governmental bodies recognize that the pressure on coastal resources is increasing, but little attempt has been made to solve the problem.

Human-Related Stresses

Several factors contribute to the depletion and degradation of fisheries resources, notably:

Pollution: Sewage, agro-industrial wastes, upon reaching the sea, cause damage to fishery resources, such as killing young fishes and other important aquatic organisms.

The Douala lagoon, Mudeka and Tiko creeks, River Mungo. Wouri, Dibamba and Sanaga, and the marine environments of Cameroon serve as centres for recreational sailing, sites for both fin and shell fisheries and as sinks for the disposal of wastes. Most of the stations where samples were taken for analysis are polluted.

The water bodies listed above support a wide array of phyto and zooplankton (These are the food for the fishes). The species composition and density of plankton are influenced not only by geographical locations but also by the water quality, which can be adversely affected by human activities.

Trace Metal Concentrations: the trace metal assayed in the water during the study were Copper (Cu), Zinc (Zn), Manganese (Mn) and Iron (Fe). These metals were also referred to as heavy metals by Reish et al, (1976): Dixit and Witcomb (1983); Mombeshora *et al.*, (1981); Oladimemeji and wade (1984); Reish (1985) and Fagade *et al.*, (1992) among other workers. In the current study, the copper values recorded ranged from 0.00-0.24ppm. The maximum allowable concentrations for fisheries and aquatic life by the EC (European Community), Canada and USSR ranged from 0.005-0.112mg 1-1. The 0.24ppm recorded, in some sampled stations were a lot higher than the specifications.

The cupric-ion concentration in some waters have been shown to be toxic to some phytoplankton species (Oben and Oben). Havens (1994) reported a reduction in the biomass of zooplankton, ciliates and flagellates by exposure to copper and the total elimination of *Microcystis aureginosa* by copper contamination. Oben and Oben (1999) reported that in most sampled stations the levels of Zinc recorded far exceeded the 0.008-0.247ppm range that adversely affected the invertebrate fauna. Values up to 1.55ppm, and 2.00ppm were recorded in Base Naval and Bonaberi bridge.

Nutrients Levels: In the area of study, increased nutrient input ranged from 0.01-0.35mg/ml (P04-P) and 0.45-1842.69mg/ml (N03-N). These may have resulted from run off; river inflow (Rivers Mungo, Dibamba, Sanaga, Ngoua and Banya stream in Douala) or sewage discharge. The nitrate concentration values obtained were astronomical compared to USSR (40mg/ml) maximum allowable concentrations for fisheries and other aquatic life (e.g. plankton; fish food). The amounts of phosphate – phosphorus (P04-P) seemed to present no danger, 0.02-0.35mg/ml, compared to 50mgl-1 advocated by the ministry of Mines, Water and Energy.

The low species diversity observed in the Douala Lagoon, Mudeka and Tiko creeks and Idenau river mouth suggest a stressed situation along the Cameroon coast. Some areas are however mores stressed than others. The immediate vicinity of the industrial zones of the Douala Lagoon, the lodging stations of the Tiko Avion beach and the area of operation of SONARA Petroleum Company thus show a high degree of pollution.

A stable tropical environment has high species diversity (Odum. 1978). The low plankton species diversity observed in the sampled stations, indicate an unstable condition occasioned by chemical discharges in the coast of Cameroon.

Inappropriate fishing techniques: Fishing techniques currently used in Cameroon target two families: *Clupeidae* (artisanal and Semi- industrial fisheries) and *Sciaenidar* (artisanal and industrial fisheries). The indiscriminate use of fishing gears results in the capture of immature fish. Fishing in nursery areas destroys eggs and juveniles. Fishing techniques involving the use of explosives and chemicals destroy the juveniles and hinder stock renewal.

Over –exploitation of stocks: Over-exploitation by industrial fishing results from the fact that the means deployed are not commensurate with the existing potential. This entails the capture of more adults than necessary, thereby hindering stock renewal. The disparity between the means deployed and existing potential can be explained by the inadequacy of knowledge about the ecology of the fishery resources (absence of inventory of exploitable stocks, evaluation and study of fish population dynamics). Stock assessment provides information about the resource abundance, composition, its age and size. Inventory and stock assessments are thus necessary to ensure efficient management of fishery resources.

Marine Fishery Management

The present fishing regulation has limits: it fails to define the different mesh sizes to be used per exploited species, to determine fishing seasons and fishing areas, to provide a clear procedure for resolving permanent conflicts between industrial and craft fisheries, and amongst fishermen themselves.

This is due to the ill-adapted legislation, poor collaboration among the different administrative units and lack of repressive measures and inefficiency of controllers. To overcome these problems in the context of Integrated Coastal Zone Management (ICM), the following actions are necessary:

- Improve, implement and enforce fishery regulations
- Update data on fishery resources
- Monitor and control fishery exploitation
- Develop inland and coastal aquaculture
- Provide infrastructure at landing sites
- Improve and institute a research and development programme
- Upgrade field supervision of fishermen and provide follow-up of their activities

The coastal forest is protected by the forestry law, but the mangrove, which is a specific type of coastal forest, is not yet protected by this law. In effect, the forest law of which the main objectives are protection of bio-diversity, does not make any specific provision for the mangrove sector. The immediate consequence of this omission is the accelerated destruction of the mangrove ecosystem which serves both as habitat and nursery ground for many fish species. Particularly fishermen in the process of constructing their homes only to name a few destroy it. Other causes of the destruction of mangrove, leading to the loss of its bio-diversity include:

- Mining and exploitation activities in the Rio-del-Rey region;
- New oil drilling around the mouth of the Sanaga River,
- Exploitation of various gravel and quarries along the coast rivers;
- Extension of port facilities and industrial zones in Douala and Bonaberi.

Climate Change Related Stresses

A number of emergent stresses associated with climate variability and change are of increasing concern to coastal zone and fishery managers, and scientists. Some of these are reviewed in this section.

- **Increasing sea temperature:** The ocean temperature changes actually long behind the air temperature changes to such a degree that even if atmospheric CO_2 concentration were stabilized today, the oceans will continue to warm for another century (Albritton *et al,* 2001). This may be a particular challenge to marine life that have relied on the thermal buffering capacity of the world's oceans to maintain a relatively consistent environment. The response of these species to rising temperatures will depend on their competitors. Most species have an optimal physiological temperature range for respiration and growth. Outside that range, individuals are energetically challenged.
- **Sea level rise**: Sea level is predicted to rise between 20 and 80cm over the next century (IPCC, 2001) This is due to both the thermal expansion of water and the melting of terrestrial ice masses. Sea level rise will affect intertidal and coastal ecosystems by inundating them with water and affecting the availability of light, as well as altering patterns of water movement both intertidally and subtidally. Patterns of the distribution and abundance of marine life are determined by salinity, light, depth and currents (Short and Neckles, 1999), rising sea levels could therefore dramatically alter communities and their composition. Stable mangrove require stable sea level. Rising sea levels changing salinity conditions and reduced sediment input due to droughts will adversely affect this ecosystem.
- **Extreme weather events**: Climate change will increase extreme weather events such as storm surges. Increased intensity and frequency of storm surges would threaten critical ecosystems along the coast.
- **Increasing atmospheric carbon dioxide:** Besides acting as a green house gas, increasing atmospheric CO_2 will result in increasing dissolved CO_2 in the water column, which in turn will cause a reduction in the pH of the ocean. As a result, less carbon

will be biologically available to calcium carbonate forming organisms (Klepas *et al;* 1999). This would weaken shells and other calcium carbonate structures, as well as slow growth rates of marine invertebrates with calcium carbonate skeletons (Hoegh-Guldberg, 1999).

Climate Change Adaptation Strategies

The management of marine fisheries is a trans-boundary issue. The countries within the Gulf of Guinea and the west coast of Africa as a whole need to design and implement the following conservation measures.

- Create reserves that contain representative ecosystem types across the "full suite" of regional biodiversity. Regional cooperation and coordination of management is required.
- Create networks of reserves which allow coverage across a gradient of biogeographic and oceanographic conditions without social, political and economic complexity of establishing a single large reserve in the region. This is important because the gradients and communities dependent on them cross political boundaries (Oslo and Dinerstein, 1990)
- Protect climate refugia at some sites. Such sites may have strong currents, upwelling, or other oceanographic features that make them less prone to thermal fluxes. Such areas may offer at least temporary refuge from warming waters resulting from climate change. These should be considered as sites for conservation efforts (Done, 2001; West and Salm, 2003). Other local physical features that may create thermal refugia include proximity to deep water, shade, high wave energy and turbidity (West and Salm, 2003)
- Effect the restoration of degraded marine sites. Unsustainable harvesting has degraded several ecosystem which are also under the stress of other human-related activities. There is need to restore and conserve such sites.
- There is an urgent need to reduce or eliminate non-climate stresses.

- Eliminate destructive fishing practices and over fishing
- Reduce pollution, including terrestrial of nutrients and pesticides; and
- Reduce damaging extraction of other resources from the ocean.

- Strategies to protect resistant and resilient populations and communities are necessary:
  - Identify those populations or communities that have endogenous factors which make then less susceptible to the effects of climate change; and
  - Maintain diverse gene pools, and natural diversity ecosystem.

Conclusion

The implementation of the above strategies requires the involvement of informed resource managers and local communities. Sufficient levels of concern and awareness of the climate change impacts on the ecosystem are necessary. Consequently stakeholders must be involved in the resilience – building process. The protection of natural resources from climate change will require a limitation of the rate and extent of anthropogenic climate change.

References

Albritton, D. and Meira, L.(2001) Techincal summary of working Group I, 3rd Assessment Report, IPCC, Cambridge University Press.

Alpha (1992). Standard methods for examination of water and waste water. American Public Health Association (ALPHA). American water works Association /Water Pollution Federation. Washington, D.C: 1193p

Berrit, G. (1969). Les eaux dessalées du Golfe de Guinée *Proceedings of the symposium on the oceanography and fisheries resources of the tropical Atlantic* FAO/OAU, Abidjan: 20-20 October 1966. p 13-22.

Dixit, S.S and Witcomb, D.(1993) Heavy metal burden in water, substrate, and macro –invertebrate body tissues of a polluted River Irwel (England). Environment Pollution (Series BO6, 161 172)

Done, T.J. (2001) Scientific principles for establishing MPA's to alleviate coral bleaching and promote recovery. In R.V. Salm and S.L Colas (eds). *Coral bleaching and marine protected areas.* Proceedings of the workshop on mitigating coral bleaching through MPA design. Bishop Museum, Honolulu

Fagade, S.O. Adebesi, A.A. Kolo, R..J; Elemi, B.F. Adeosum A.A. and Ja'a afru (1992) Urbanization and degradation of aquatic resources. The urban experience. Paper presented at the National conference on Conservation of Aquatic Resources, Organised by a National Advisory Committee on Conservation of Renewable Resources. Conservation council, Abuja (11th -14th May 1992) p. 1-18.

FAO (1971) Consultation on the conservation of fishery resource and the control of fishing in Africa. Casablanca: 20-26 May 1971, FAO Rome. 44p

Havens, K.E. (1994). Structural and functional responses of a fresh water plankton community to acute Copper stress. *Environmental Pollution, Vol. 86,* p. 259-266

Ibe, C.A. Folack, J.Mbame; Bokwe, A. and Tega, A.(1999) Cameroon Profile, Large Marine Ecosystem Project for the Gulf of Guinea. Printed by Ade Graphics Industries, Yaounde, Cameroon, p. 34-40.

IPCC (2001) Climate change 2001: The scientific basis. Cambridge University Press.

Jeje, C.Y.and Fenando, C.H. (1992). Zooplankton associations in the middle Niger-Sokoto Basin (Nigeria: West Africa) Int. Revue. ges. *Hydrobiol.* Vol. 77, p. 237-253.

Kleypas, J., Buddemeier, D., Archer, D., Langdon, C et al (1999) Geochemical consequences of increased carbon dioxide on coral reefs. *Science, Vol. 234,* p. 118-120.

Kolesnikov, A.G. et al (1971) Lomonosor current in the Gulf of Guinea. *Oceanography Vol 11,* p. 311-315

Mazeika, P. (1968) Eastward flow within the south Equatorial current in the eastern South Atlantic. *Journ. Geophs Res. Vol. 73,* p. 5819-5828.

Mombeshora, G. Ajayi, S.O. and Osibanji, O. (1981) Pollution studies on Nigerian rivers: Toxic heavy metal status of surface water in Ibadan city. *Environment International Vol. 5,* p. 49-53.

Morin, S.(1982) Notes hydrographie et hydrologie Camerounaises. *Travaux et Doument de Geographie.* University of Yaounde. 64p.

Needham, J.G and Needham PR. (1975). A guide to the study of fresh water biology. Holder Day publ. San Francisco, 5th ed.

Ngwa, J.A (1979) a new geography of Cameroon. Longman Group Ltd. London 151p.

Noss, R. (2001) Beyond Kyoto: Forest management in a time of rapid climate change. *Conservation Biology, Vol. 15,* p. 578-590.

Oben P.M.; and oben B.O. (1999) Plankton survey. Final Report on Plankton Survey in the National waters prepared for the United National Industrial Development in the Gulf of Guinea. Large Marine Ecosystem Project. 45 p.

Odum, E.P. (1971) Fundamentals of Ecology, 3rd W.B. Saunders co, Philadelphia (1985): Trends expected in stressed Ecosystems. *Bioscience , Vol. 35,* p. 149-422

Oladmeji. A.A and Wade, J.W. (1984) Effects of effluents from the sewage treatment plant on the aquatic organisms. *Water, Air and Pollution. Vol. 22*, p. -7

Olson, D. and Dinerstein, E. (1998) The global 200: a representation approach to conserving the earth's most biologically valuable eco-regions. *Conservation Biology Vol. 12*, p. 502-515.

Pertroth, I. (1969) The distribution of water type structure in the first 300 feet of the Equatorial Atlantic. Proceeding of the symposium on Oceanography and fishery resources of the tropical Atlantic, Abidjan, UNESCO/FAO/OAU: 20-28 October 1961, p. 185-191

Reish, D.J. (1985) The use of *Polychaetous annelid Neanthes arenceodentata* as a laboratory experimental animal. Tethys 1 (3-4), 335.

Rinkel, M., Sund, P., Neumann, G (1966) The location of the termination area of the Equatorial under-current in the Gulf of Guinea based on observations during Equalant III. *Journ. Geophys. Res Vol. 71*, p. 3893-3903

Short, F. and Neckles, H.(1999) The effects of global change on sea grasses. *Aquatic Botany, Vol. 63*, p. 169-196.

Ssentongo, G.W . (1979) Marine Fisheries in Africa. *Proceedings of 6th workshop in Aquatic pollution in relation to the protection of living Resources.* FAO/SIDA: Nairobi – Mombassa. p. 19-38

West, J and Salm, R. (2003) Resistance and resilience to coral bleaching: implications to coral reed conservation and management. *Conservation Biology, Vol. 17*, p. 956-967.

Chapter Seven

Climate Change and Management of Riverine Ecosystems

Summary

Global warming has distorted rainfall and temperature patterns across the globe and this is equally being felt in Cameroon. It has been observed to reduce rainfall, increase the frequency of extreme events such as droughts and floods, and therefore surface and groundwater resources in the main watersheds of the country. This chapter uses a combination of field observations and secondary data to overview the underlying physical processes of freshwater systems in Cameroon that are most vulnerable to change and to determine how these changes might affect aquatic life. It concludes that aquatic life in river systems are sensitive not only to water temperature, volume and flow, but also to variability in these factors. The variety of changes to be displayed by rivers in response to global climate change are anticipated and data and strategies for designing and managing river systems that will allow them to withstand and to adapt to climate change are provided based on observed and predicted effects of global climate change. Finally, general suggestions are presented for designing and managing river fisheries to increase resistance and resilience to climate change.

Keywords: Climate change, physical processes of freshwater systems, river fisheries, resistance, resilience, climate change impacts, management.

Introduction

Freshwater ecosystems have been critical to sustaining life and establishing civilizations throughout history. Man relies on them for drinking water, agriculture, transportation, energy production, industrial processes, waste disposal and extraction of fish and other products. Freshwater fishes account for over 25% of the living vertebrate species. (McAllister *et al*; 1997). Temperature and rainfall are the main conditions which sustain the growth and development of biodiversity. Because freshwater ecosystems are sensitive not only to water temperature, volume, and flow but to variability in these factors, rivers, lakes and wetlands are expected to display a variety of changes in response to global climate change.

This chapter uses a combination field observations and secondary data to overview the underlying physical processes of freshwater systems in Cameroon that are most vulnerable to change and to determine how these changes might affect aquatic life. The purpose is to provide data and strategies for the design and management of river systems in order to enable them to withstand and/or adapt to climate change.

Ecological and Socio-Economic Setting

Cameroon has a total land area of about 473,440km^2 with a population of 17 million and a growth rate of 2 percent. Agriculture employs 70 per cent of labour force and contributes 44 percent to the Gross Domestic Product.

The country has three climate regions (Figure 1). These are the humid-wet equatorial region in the South, a montane equatorial climate in the West, and a dry semi-arid climate in the North. Rainfall decreases from South to North. Figure 1 presents the actual rainfall and temperature regimes. There are high year-round temperatures and the weather is controlled by equatorial and tropical air masses. It has two major seasons, a dry season and a wet season. Most of the rain occurs between April and October. Rainfall is variable and unreliable (Ngakfumbe, 2001) and floods and droughts occur frequently.

The Southern part of the country has an average temperature of 25^{0}C while the northern regions between 27^{0}C and 34^{0}C. The northern parts have seasonal rivers. These dry up during the dry season. The southern regions have abundant surface water resources. These drain into the Congo basin; Niger basin and the Gulf of Guinea while the northern rivers but for the Benue River drain into the Chad basin.

The freshwater life is extremely varied. There are about 550 species of freshwater fish composed of over 47 families and 148 types. The most important are the Cyprinidae (50 species). The Circhlidae family is well represented by *Tilapia* spp. The Mormymidae include 50 different species. The Synodontae are composed of 30 species and present in almost all rivers. The Citharindae family is represented by 16 species. Table 1 presents an inventory of fish in the Manyu River and its tributaries. A case study was made of the Manyu River basin (Figure 3) in order to obtain data on the ecological and socio-economic setting of the study.

Fishing contributes to local diets and the trading economy of villages within and around riverine areas (du Feu, 2002). In the Manyu River, fish breeding and migration occur in the small tributaries during the dry season (October to December). This period marks the time when hatching produced in the preceding rainy season use the flood streams as breeding grounds before returning to the main rivers. It is during the same period, fish fencing, traps, cast nets, and cross-over nets that cause high mortality to juvenile fish are frequently operated.

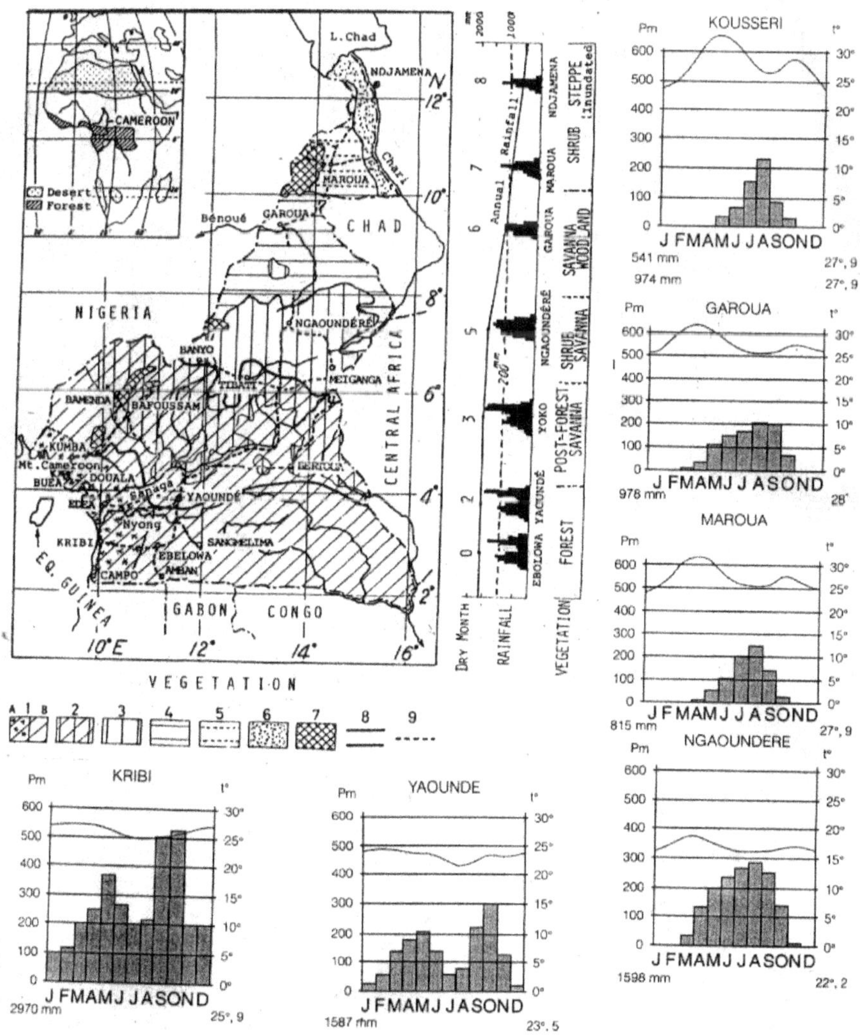

Climate, vegetation and physiography of Cameroon. Vegetation (after LETOUZEY, 1968, 1979) Dense humid evergreen forest, including mangrove, 1B: Dense humid semi-deciduous forest, 2: Post – Forest savannah (grass and shrub savannas), 3: Shrub and grass savannas of Adamaoua Plateau, 4: Tree savannah, savannah woodland and dry woodland (Sudanese zone), 5: Tree and shrub steppes with thorn bushes (Sahelian zone), 6: do, inundated periodically, 7: Montane and subalpine formation, 8: Zone of ancient and recent destruction of dense humid forest, 9: Main survey routes of the 1980/81 season fieldwork.

Figure 1: Climate, vegetation and hydrography of Cameroon.

Table 1: List of species in Manyu River System (After Mdaihli *et al*; 2003)

Family	Species	English name	Pidgin name
Bagridae	*Auchenoglanis bisculatus*	Catfish	
Bagridae	*Bagrus docmak*	Silver catfish	Male belly up
Bagridae	*Bagrus flamentosus*	Silver catfish	Male belly up
Bagridae	*Chrysichthys nigrodigitatus*	Catfish	Belly up (female)
Centropomidae	*Late niloticus*	Nile perch	Gouper
Characidae	*Brycinus brevis*	African tetras	Mbanga
Characidae	*Brycinus longipennis*	African tetras	Mbanga
Characidae	*Hydrocynus brevis*	Tiger fish	Dog fish
Characidae			Mbanga
Characidae	*Micralestes elongates*		Mbanga
Cichlidae	*Micralestes humilis*	Tilapia	Bone back
Cichlidae	*Pelvicachromis pulcher*	Tilapia	Bone back
	Pelvicachromis taeniatus		
Cichlidae		Tilapia	Bone back
Cichlidae	*Sarotherodon galilaeus*	Tilapia	Bone back
Citharinidae	*Tilapia mariae*	Moon fish	Sand leaf fish
	Citharinus citharus		
Citharinidae		Moon fish	Sand leaf fish
Clariidae	*Citharinus latus*	Catfish	Mudfish
Clariidae	*Clarias agboyiensis*	Catfish	Mudfish
Clariidae	*Clarias anguillaris*	Catfish	Mudfish
	Heterobranchus bidorsalis		
Cyprinidae		Barbs	Mbanga
	Barbus bynni occidentalis		
Cyprinidae		Barbs	Mbanga
Cyprinidae	*Barbus lagoensis*	African carps	Craw-craw nose
Cyprinidae	*Labeo parvus*	African carps	Craw-craw nose

Cyprinidae	*Labeo senegalensis*		Aeroplane fish
Dasyatidae	*Raiamas nigeriensis*	Ray	Cover pot
	Dasyatis garouaensis		
Denticiptidae		Grass-eaters	Mbanga fish
Distichodontidae	*Denticeps clupeoides*	Grass-eaters	
Distichodontidae	*Distichodus engycephalus*	African pike	
Hepsetidae	*Ichthyborus monodi*	Electric fish	Dog fish
Malapteruidae	*Hepsetus odoe*	Spiny-eel	Electric fish
Mastacembelidae	*Malapterurus electricus*		Snake fish
	Aethiomastacembelus	Spiny-eel	
Mastacembelidae	*nigromarginatus*	Catfish	Snake fish
Mochokidae			Knock-a-knock
	Caecomastacembelus decorsei	Trunkfish	Postmaster
Mormyridae	*Brachysynodontis batensoda*		Elephant fish
		Elephant fish	
Mormyridae	*Campylomormyrus tamandua*	Trunkfish	Elephant fish
Mormyridae		Trunkfish	Elephant fish
Mormyridae	*Gnathonemus petersii*		Elephant fish
	Hippopotamyrus pictus	Trunkfish	
Mormyridae	*Hippopotamyrus psittacus*	Trunkfish	Elephant fish
Mormyridae		Trunkfish	Elephant fish
Mormyridae	*Marcusenius cyprinoides*		Elephant fish
	Mormyrops oudoti		
Mormyridae	*Mormyrus macrophthalmus*	Trunkfish	Elephant fish
Mormyridae		Trunkfish	Elephant fish
Mormyridae	*Mormyrus rume*	Trunkfish	Elephant fish
Nandidae	*Mormyrus spp.*		Grouper
	Mormyrus tapirus	Feather back	
Notopteridae	*Polycentropsis abbreviate*	African knife-fish	Canada planty
Notopteridae		Bony tongue	Canada planty
Osteoglosidae	*Papyrocranus afer*	Blood fish	
Phractolaemidae	*Notopterus nigri*	Bichir/Thorny-	Mudskip
Polypteridae	*Heterotis niloticus*	eel	Snake fish
Schilbeidae	*Phractolaemus ansorgii*	Butterfish	Female belly up
Schilbeidae	*Polypterus ansorgii*	Butterfish	Female belly up
Tetraodontidae	*Schilbe brevianalis*	Puffer fish	Football fish
Tetraodontidae	*Schilbe intermedius*	Puffer fish	Football fish
	Tetradon lineatus		
	Tetradon pustulatus		

Source: Mdaihli, 2003

Migratory *Labeo Spp*. Make up 10% of all recorded catches in the Manyu River. This level is probably much higher when one includes catches during annual migration. The commercial importance of these fish presents a case for controlling the fishing effort during their migration. Within the rainforest streams there is the large breeding of some species. In the Manyu River the most common is the migration of *Labeo batesii, Hemigrammopetersius brevidorsalis*, and *Barbus sp*. Mdaihli *et al*, 2003 recognize 166 species from 15 orders in Manyu River. Species include those form the lower Guinean ichthyo-faunal province and the upper Guinean, Nilo-Sudanian, and Zarian provinces. The Manyu River/Cross – River has more fish species than any other comparable West African river basin and presents a rich Ichthyofauna (Teugels *et al*; 1992). Small villages depend on these for a livelihood. The fish is smoked and sold on local and regional markets.

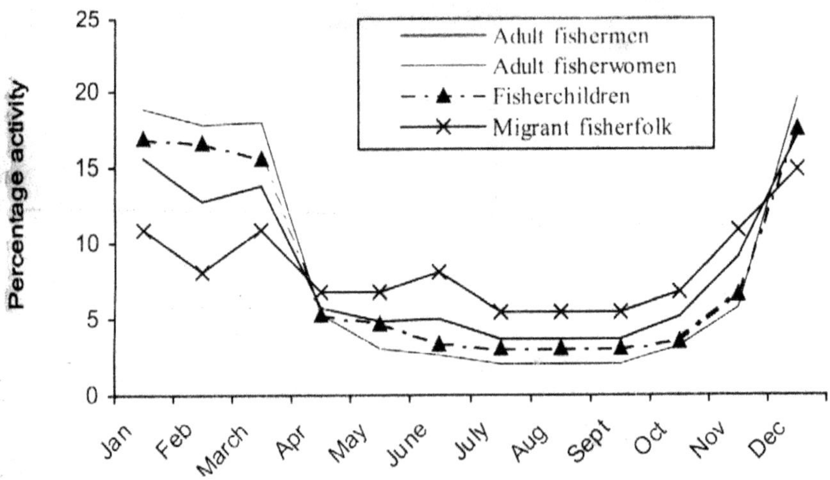

Figure 2: Seasonality of fishing by fisherfolk South of Takamanda in Manyu River basin (after Mdaihli, 2003).

About 40% of the adult population in the riparian areas of Manyu fish, mostly for home consumption. Fish that are sold are either fresh (50%) or smoked (50%). Figure 2 presents population of fisherfolk per season in Takamanda village. Fishing takes place all year round with peak activity extending from the end of the rainy season. (December) into the short dry season in March. At this time the water currents are slow and the river /

stream levels less variable than during the wet season. Fish also become more concentrated as the river level drops, making them easier to catch. There is a rapid rise in fishing activity from October to December after the wet season when streams and rivers are emptying. During this period a large number of juvenile fish that hatched during the rains and use flooded areas and streams as nursery grounds are caught-primarily through the use of fish fences, traps, cast-nets, and cross-over nets. However, fishing for juveniles that have not had a chance to reproduce reduces the overall annual fish production of the water bodies. For the Takamanda villages Mdaihli *et al.* (2003) estimate an annual yield of 1.066tons worth approximately FCFA 400 million.

Figure 3 presents the main freshwater fisheries in Cameroon. The fish production in Cameroon is declining to the advantage of fish imports (Figure 4). The inland waters can be divided into lacustrine systems and riverine systems with associated swamps, river mouths and coastal lagoons. In all these inland aquatic systems, there is a high degree of spatial heterogeneity (the same species are distributed in many geographically isolated lake basins and rivers). But it is the streams and rivers which have maintained this heterogeneity. These fisheries are very susceptible to depletion due to the high fishing intensity immediately after the rainy season.

Fish production can be increased if proper fisheries management and sustainable development policies are adopted. Increased fishery research and greater awareness of the resource potential will lead to the solution of these problems. But some management and development problems might arise from the number of conflicts due to the multi-purpose use of these inland waters and the vulnerability of these fisheries to climate change.

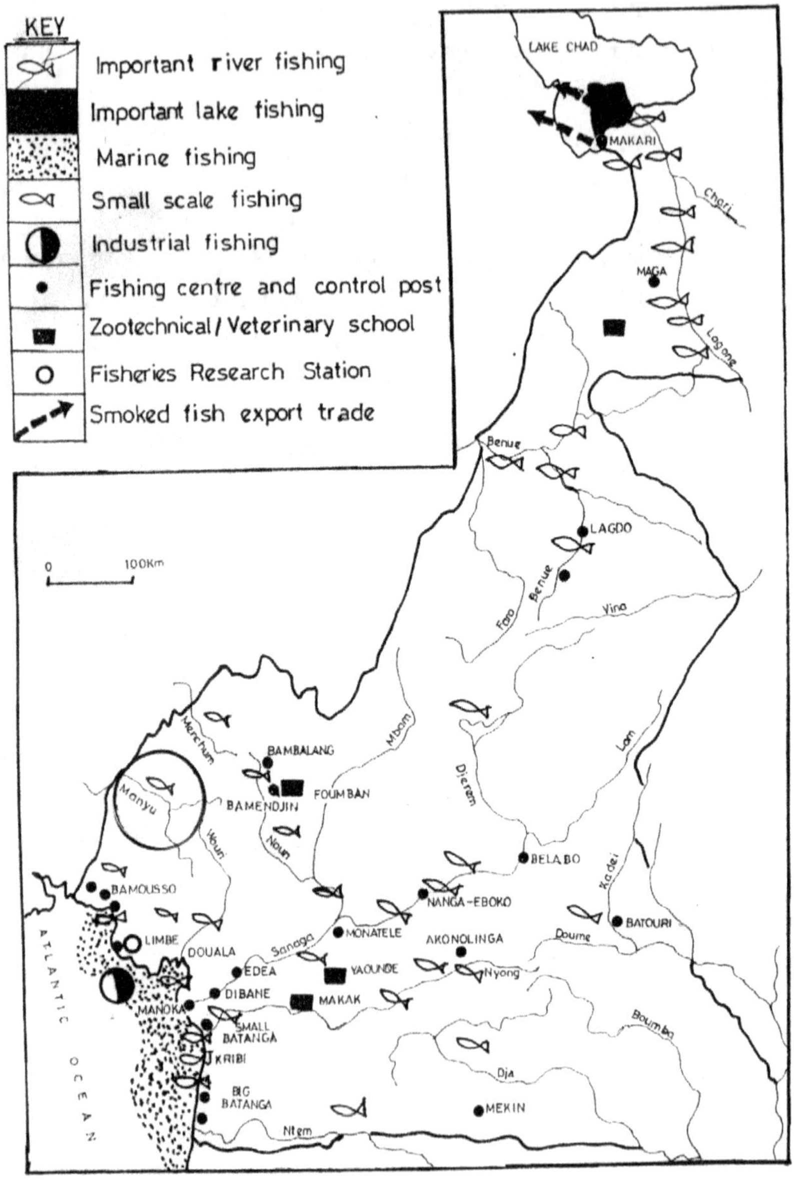

Figure 3: Location of the main freshwater river fisheries in Cameroon.

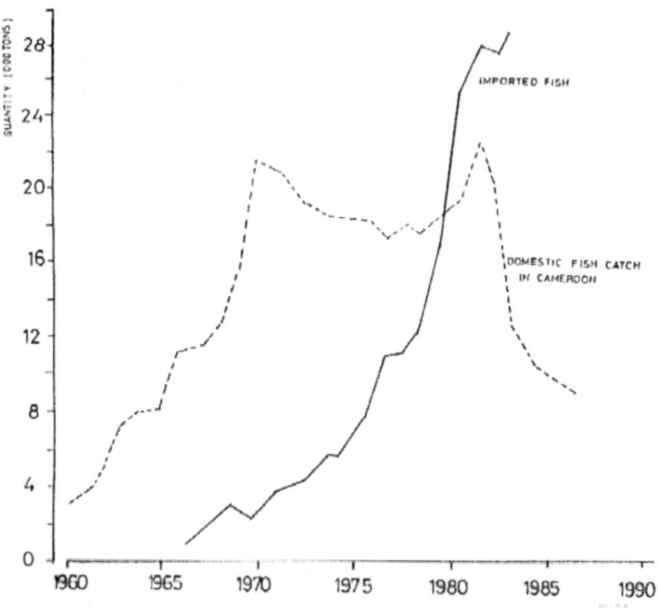

Figure 4: Evolution of fish imports and domestic fish production in Cameroon.

The Concept of Stability and Resilience

With a wide geographical distribution of certain fish species in the various aquatic systems and in various eco-climatic regions, the available fish species will display differential rates of survival and population growth depending on the environmental factors controlling stability and resilience. There is less stability in temporary fluctuating lakes. For example, Lake Chad has a rate of evaporation of 7000mm/year. It is an inland drainage lake which is increasingly shrinking in size because of much evaporation due to high temperatures. The increased salinity of the lake has resulted in more blue green algae production. The seasonal rivers of the northern regions, that is, the Benue basin and the Chad basin are also characterized by fluctuating water levels and low stability.

On the other hand, the lakes of the south Cameroon plateau and the West Cameroon Highlands show less variation in physical and chemical characteristics. However, global warming and increased precipitation is expected to provoke fluctuations in water levels and low stability. Riverine systems will similarly suffer from climate change induced droughts and floods. Both in the stable and unstable aquatic systems, fish species do persist and display varying degrees of resilience, that is, response to external perturbations. It is apparent that the available space and food resources determine the niche space and abundance of fish species in various freshwater systems. Precipitation and temperature changes will therefore affect the stability of freshwater fisheries.

Temperature and Precipitation Changes

Ayonghe (2001) showed that the net trend per decade in Cameroon based on data between 1930 and 2000 is a temperature increase of 0.14^{0}C. The highest rates of temperature increase per decade were at Banyo (0.24^{0}C), Maroua (0.23^{0}C), Kribi (0.32^{0}C) and Bertoua (0.21^{0}C). The lowest rates were at Ngaoundere (0.06^{0}C) and Mamfe (0.04^{0}C). In the same study plots of total amount of rainfall monitored at 11 stations showed decreasing trends in 6 stations. High rainfall was evident from 1951 to 1967, 1977 to 1980, and 1989 to 1995, while low rainfall was evident from 1930 to 1950, 1968 to 1976, and 1981 to 1988. In the drier regions of North Cameroon the data between 1960 and 1990 were in agreement with similar studies in the north-eastern arid zone of Nigeria where Carter and Alkali (1996) reported rainfall as decreasing by 8mm per year. In the Nyong River basin, rainfall variability is between plus 10% and minus 5%. Extreme events provoke a variability of plus 20%. The net changes in temperature across the entire country from 1930 to 1935 were 0.91^{0}C, which is equivalent to 0.14^{0}C per decade. Similarly, the net change in the total amount of rainfall was minus 282mm, equivalent to minus 43mm per decade. The net change in the annual number of rainy days minus 7 days over the same period. When projected to the year 2060, these net changes give a temperature increase of 1.8^{0}C and rainfall decrease of 559mm, and 16 days respectively. The changes in mean number of rainy days are presented in table 2.

Table 2: Mean total number of rainy days computed from trend lines.

Stations	**1940**	**1960**	**1980**	**1995**	**2030**	**2060**
Garoua	82.1	79.7	77.2	75.4	71.1	67.4
Ngaoundere	145.9	146.8	147.7	1488.4	150.0	151.3
Mamfe	248.4	222.4	196.5	177.0	131.5	92.6
Bamenda	213.5	204.3	195.0	188.0	171.8	157.9
Banyo	138.4	151.6	164.9	174.8	198.0	217.8
Maroua	72.8	71.8	70.8	70.1	68.4	66.9
Douala	142.8	229.6	216.3	206.4	183.3	163.4
Kribi	209.0	205.1	201.1	198.2	191.3	185.4
Yaounde	148.9	149.1	150.5	151.2	152.9	154.3
Ebolowa	156.4	163.8	171.2	176.8	189.7	200.9
Bertoua	123.1	127.4	131.8	135.0	142.6	149.1
Mean	161.9	159.1	156.6	154.7	150.0	146.1

Source: Ayonghe, 2001

The effects of global warming on air and water temperatures are likely to be far more complicated than a gradual increase in average air temperatures. Both observational studies and models of future climate change suggest that there will be more variable temperatures, a greater frequency of extreme temperature events and more hot dry season days. They also predict a change in rainfall variability (Easterling *et al*; 2000), an increase in the frequency of extreme rainfall due to an increase in the intensity of tropical storms (Anderson *et al*; 2002). These will result in increased droughts and floods.

Temperature change alone is known to affect a range of physiological processes and life history traits. Higher ambient temperatures increase the metabolic demands of many animals. The energy requirements of aquatic life will therefore increase (McDonald *et al*; 1996). The effects of higher metabolism on growth may depend on food availability. In zooplankton with an adequate food supply, increased food supply lead to a dramatic rise in feeding, assimilation, growth and reproduction rates (Schindler, 1968), and local species richness can increase as a result (Stemberger *et al*; 1996). Increased temperatures can also result in the rise in the frequency of toxic algal outbreaks, and in their toxicity to other aquatic animals (Hallengraeff, 1993). Increased temperatures have been shown to cause a reduction in the body size at a developmental stage in over 90% aquatic animals (Atkinson,

1995) and to determine the sex of some offsprings. This skews sex ratios for most amphibians (Janzen, 1994).

Climate warming is likely to alter the composition of aquatic communities, as different species will have different thermal tolerances and interactions between species may intensify as a result of reduced resources and habitat availability (Rhee and Gothan, 1981). Where water levels in rivers decrease as a result of drought, biotic interactions including human overexploitation may intensify as a result of increased densities of aquatic life (Feminella and Resh, 1990).

In the tropics water temperature is mainly regulated by shade and rainfall. The rainy and dry seasons lead to large, predicated, seasonal variations in rainfall and annual flooding of adjacent forests and grasslands, which provide abundant food and breeding grounds for fish. Thus, the life histories of river fish will be more strongly affected by changes in water level, than by changes in temperature.

Water Quantity and Flow Changes

Life history traits are intricately linked with water quantity and seasonal flow in numerous aquatic animals. Many fish in the tropics undergo feeding and spawning migration of several thousand kilometres that are dependent on predictable, seasonal flooding events (Junk, 2002), and extreme flow levels may be necessary for maintaining populations of a number of other species (Poff *et al*; 1997). On the other hand, intense flooding can scour stream beds, displacing organic matter, bottom feeding organisms, and small fish, and substantial increases in flood frequency could cause a shift in species composition, possibly eliminating many species.

Flow regime is therefore a critical component of river ecosystem. Mean flow may increase or decrease depending on changes in temperature and rainfall. Many rivers will experience altered timing or duration of high and low flows due to changes in seasonal variability of rainfall and frequency of extreme rainfall events. Extreme flooding events and landslides could remove important woody debris from rivers and destabilize river channels. Where rainfall reduces, stream flow volume may also decrease, and reductions in runoff will lower the concentration of organic matter in rivers. A reduction in natural flooding events could eliminate many of the beneficial effects of seasonal flooding, such as creating flood plain habitats and determining channel form. Changes in flood plain dynamics will directly affect the populations and fishery yield, as growth rates and overall fish catch is correlated with the area of the flood plain (Meisner and Shuter, 1992).

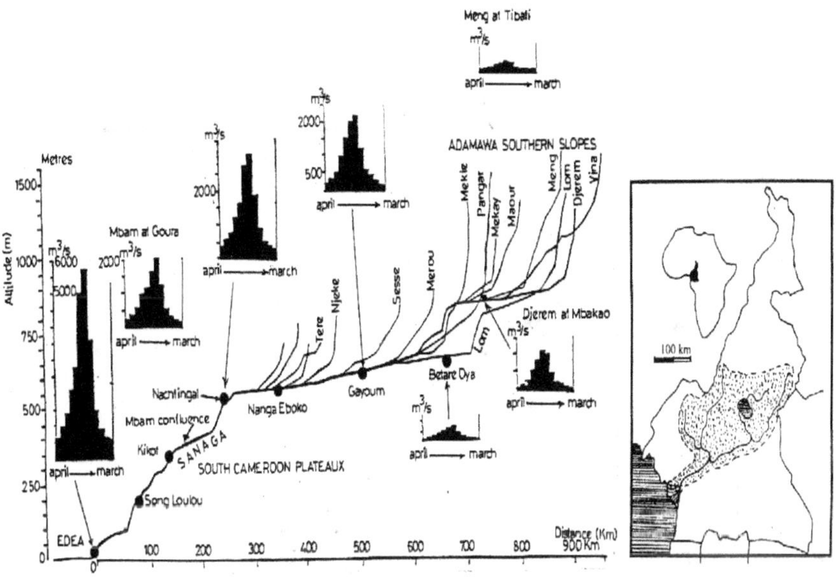

Figure 5: River regimes and longitudinal profiles in the Sanaga River basin

(After Morin, 1982).

The hydrological regimes of the country are strongly influenced by the climate. Equatorial regimes in the South Cameroon Plateaux typically have two flood peaks and two low stages. These are the Lobe and Lokoundje Rivers. Further inland in the centre region are tropical transitional regimes with a single flood peak. Most rivers of the Sanaga River basin (Figure 5) have this regime. In the northern regions the regimes are characterized by marked seasonality. The rivers of the Chad and Benue basin completely dry up during the dry season.

Although precipitation is one of the main factors determining water availability and flow, other factors such as evaporation, soil moisture and groundwater recharge are also critical. Evaporation is generally expected to increase due to increasing mean temperatures. Although climate change is likely to lead to some changes in groundwater recharge, freshwater systems that primarily receive input from groundwater such as those with equatorial regimes will experience smaller changes. Rivers with tropical regimes solely dependent on rainfall will experience higher variations.

Water Quality Changes and Fluvial Processes

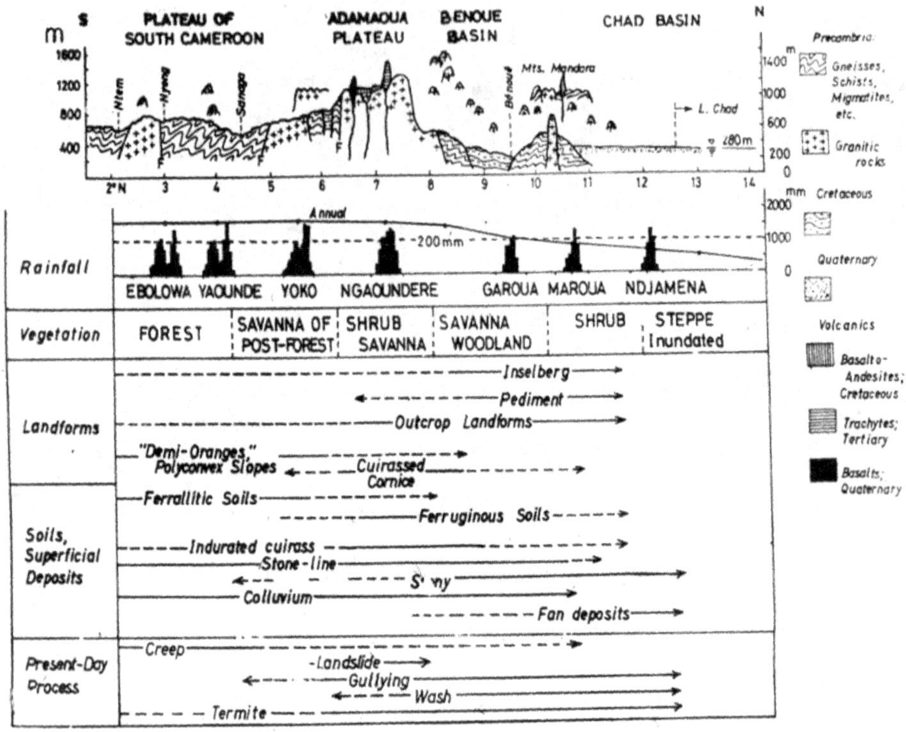

Figure 6. South – North variation of landscape, vegetation, rainfall, landforms, superficial materials and processes in Cameroon (After Kadomura, 1986).

Figure 6 presents the South – North variation of landscape, landforms, superficial depositis and geomorphic processes on drainage basin slopes in Cameroon. The supply of nutrients in rivers is affected by the vegetation, turbidity, the type of soil in the drainage basin and the degree of leaching. In Cameroon, the rainfall, vegetation, topography of the drainage basin and geology of the rocks vary considerably. Streams and rivers obtain most of the maintenance energy outside the riverine systems. Also characteristic of rivers is the constant downstream displacement of inorganic and organic materials which result in longitudinal variation in biogeochemical cycles. This means that the trophic relationships and the interaction of fish species vary longitudinally. Consequently, fish production varies longitudinally with higher production in the middle and lower reaches of rivers where swamps and flood plains occur (Ssentongo, 1979).

Heavy seasonal rainfall events lead to high sediment input into rivers and dangerous levels of pesticide runoff from agricultural lands in drainage basins could lead to water quality problems. The drainage basin geomorphic processes produce the suspended sediment loads presented in figures 7 and 8. Extreme droughts and rainfall events induced by climate change could therefore lead to further water quality problems. The sedimentation of river ports such as the Mbalmayo port on the Nyong River, and the Sanaga River lower reaches are evidences of climate change.

Many climate change models predict that extreme water flow events, such as floods are likely to increase, due to heavier individual rainfall events (Reynard *et al.* 1998). Such events will therefore adversely affect riverine fisheries. Sediments, pesticides, animal wastes, urban and industrial wastes will be washed into rivers. These will become polluted if the rate of discharge of these wastes exceeds the absorptive rate of that water body. Some models also predict seasonal shifts in peak flooding seasons (Saelthun *et al.*, 1998).

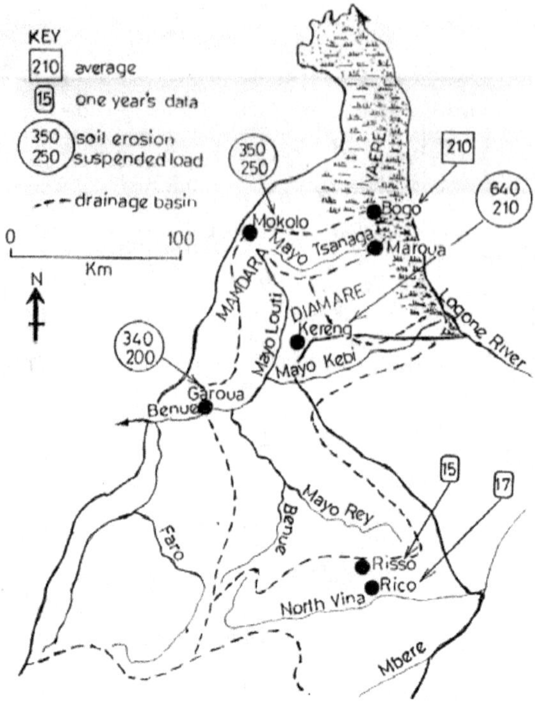

Figure 7: Suspended sediment load in rivers of the Benue and Chad basins (after Olivry, 1978) in tons/km^2/year.

The input of chemicals, sediments, organic matter, nutrients and pollutants to freshwater ecosystems are like to be affected by climate change. The forest regions are characterized by deep weathering and leaching of bases and sesquioxides released by hydrolysis. In the Sahel and Savanna regions of the north surface wash processes and gullying are active. In the West Cameroon Highlands gullying, runoff erosion and landslides are common. Both simulations and direct observations indicate that intense storm events following prolonged dry periods can lead to increased flushing of sediments or nitrates into water bodies (Arnell *et al*; 2001).

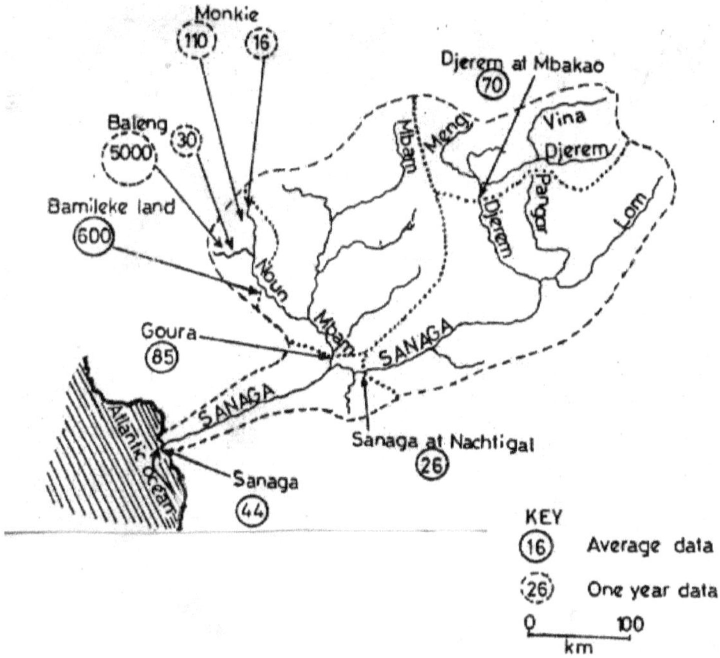

Figure 8: Suspended sediment load in the Sanaga River basin (after Olivry, 1978) in tons/km^2/year.

Rising temperatures and changes in rainfall are likely to cause changes in the vegetation in watersheds. These changes may affect the supply of organic matter to freshwater systems, the characteristics of runoff entering the system, (nutrient concentrations and sediment load). Rising temperatures and reduction in rainfall may cause an increase in bush fire incidences in the savannah, and Sudano - Sahelian regions. This could lead to increased nutrient input from burnt vegetation leading to eutrophication, increased sediment load and reduced input of organic matter. Decreased water levels as a result of droughts could, on the other hand, lead to concentration of pollutants from point sources (Schindler, 2001).

Planning and Management Framework

The persistence of individual species or communities in the face of climate change depends on how resistant and resilient they are (Noss, 2001). Because there is still a great deal of uncertainty associated with climate change predictions, interactions between the biotic and abiotic components of freshwater ecosystems can be complex and non-intuitive. Focusing on increasing system resistance and resilience is the main approach. One cannot plan for a specific set of predicted changes. The following measures can be adopted in the management of freshwater ecosystems to withstand climate change:

- ***Preservation of habitat heterogeneity and biodiversity*:** Both species and habitat diversity increase resistance and resilience to climate change. This is because diversity provides a greater range of stress tolerance and adaptive options (Chapin *et al.* 1997). High biodiversity areas may also become important as sources for colonizing new areas as the effects of climate change become more severe (De Groot and Ketner, 1994).
- ***Protection of physical features rather than individual species*:** Aquatic ecosystem function is determined by basic physical features such as water flow, channel morphology, and nutrient balance, rather than by species assemblages (Moss, 2000). Protecting flow patterns, water quality and water quantity will promote the protection of biodiversity in freshwater habitats (Abell *et al;* 2002).
- ***Preservation of habitat connectivity to allow access to migration routes and thermal refugia*:** This can help preserve flow regimes, promote ecological integrity, and allow migratory fish to move between different habitats at various life history stages (Abell *et al;* 2002). It can also provide fish with access to thermal refugia, or allow them to migrate to more suitable habitats.
- ***Protection of rivers from human pressures and exotic species*:** Stresses that can reduce species diversity must be minimized. This is because stressed systems display reduced resistance and resilience to change (Noss, 2001).
- ***Promote the management of entire watersheds and regulate extractive water use:*** Freshwater ecosystems are intricately connected to their drainage basins, and downstream river segments can be extremely sensitive to distant, upstream disturbances such as deforestation, agriculture and other activities that alter water quality, quantity and temperature (Moss, 2000).
- ***Undertake active interventions in response to climate change:*** The objective is to ameliorate the effects of climate change

or to directly protect vulnerable species. Inter-basin transfer of water has been suggested (Pringle, 2000). Flow conditions below dams such as the Mbakao, Bamendjim, Lagdo and Songlou-lou dams can be manipulated to relieve some of the negative effects of climate change, for example, by releasing bursts of water to stimulate flooding events (Middleton, 1999).

- ***Employment of adaptive management strategies to maintain flexibility:*** Aspects of climate change are unpredictable. It is therefore crucial to maintain flexible conservation goals and strategies. Management can allow for and even assist in the adaptation of species and communities. The aim should be to protect ecosystem processes (Markham and Malcolm, 1996).

Applying the above set of goals will result in some conflicts. There is a need to evaluate how best conflicting goals can be met. The adaptation strategy for each river basin must therefore seek to (Anon, 2003):

- Characterise and monitor species /systems that are most vulnerable to climate change to determine the scale of conservation needs.
- Develop strategies for protecting freshwater ecosystems while dealing with increasing water needs.
- Perform integrated river basin management to buffer systems against climate change and provide a basis for long-term conservation.
- Protect entire freshwater ecoregions to protect the ecological and evolutionary driving forces of biodiversity.

References

Abell, R., Thieme, M.; Dinerstein, E.; and Olson, D (2002) A source book for conducting biological assessments in developing biodiversity visions for ecoregion conservation. *Freshwater Eco regions, Vol. 2.* WWF, Washington D.C. 201p.

Anderson, D.; Overpeck, J and Gupta, A. (2002) Increase in the Asian Southwest monsoon during the past four centuries. *Science Vol. 297* p. 596 – 599

Arnell, N.; and Lui, C. (2001) Hydrology and water resources. In: J. Mclarthly, N. Canziana, D. Leary, Dokeen and K. White (eds). *Climate change 2001: impacts, adaptation, and vulnerability.* Intergovernmental Panel on Climate Change, Cambridge University Press, p. 191 – 233.

Atkinson, D. (1995) Effects of temperature on the size of aquatic ectotherms: exceptions to the general rule. *Journal of Thermal Biology, Vol. 20*, p. 61 – 74

Anon. (2003) user's manual for building resistance and resilience to climate change in Natural Systems. WWF Climate Change Programme, Washington D.C. p. 175 – 214.

Ayonghe, S. (2001) A quantitative evaluation of global warming and precipitation in Cameroon from 1930 to 1995 and projections to 2060: effects on the environment and water resources. In: E. B. Eze and C. M. Lambi (eds). *Readings in geography*. Unique Printers, Bamenda, p.142 – 155.

Carter, R. and Akali, A. (1996) Shallow groundwater in the North-eastern arid zone of Nigeria. *Quarterly Journal of Engineering Geology, Vol. 29*, p. 341 – 355.

Chapin, F.S.; Walker, B.; Hobbs, R.; Hooper, D; Lawton. J; Sala, O.; and Tilman, D. (1997) Biotic control oven the functioning of ecosystems. *Science, vol. 277*, p. 500 – 504.

Du Feu, T. (2001) Fish and fisheries in the Southern zone of Takamanda Forest Reserve, South West Cameroon. PROFA Report, Mamfe.

De Groot, R. and Ketner, P. (1994) Sensitivity of NW European species and ecosystems to climate change and some implications for nature conservation and management. In: J. Pernetta, D. Leemans, D. Elder and S. Humphrey (eds). *Impacts of climate change on ecosystems and species: implications for protected areas.* IUCN, Gland, p. 28 – 53.

Easterling, D.; Meehl, G.; Parmesan, C.; Changnon, S.; karl, T.; Mearns, L. (2000) climate extremes: observations, modeling and impacts. *Science, vol. 289,* p. 2068 – 2074.

Feminella, J. and Resh, V. (1990) Hydrologic influences, disturbance, and interspecific competition in a stream caddisfly population. *Ecology, vol. 71*, p. 2083 – 2094.

Hallengraeff, G. (1993) A review of toxic algal blooms and their apparent global increase. *Phycologia, vol. 32*, p. 79 – 99.

Janzen, F. (1994) Climate change and temperature dependent sex determinants in reptiles. *National Academy of Sciences, vol. 91,* p. 7487 – 7490.

Junk, W. (2002) Long-term environmental trends and future of tropical wetlands. *Environmental conservation, vol. 29*, p. 414 – 435.

Kadomura, h. (1986) Geomorphology and environmental changes in tropical Africa. Case studies in Cameroon and Kenya, Sapporo, Japan, 299p.

Letouzey, R. (1968) Etude phytogeographie du Cameroun Paul. Lechavalier Paris, 511p.

Letouzey, R. (1979) Vegetation. In: J. Loung (ed) *Atlas of the United Republic of Cameroon*, Jeune Afrique, Paris, p. 20 – 24.

Markham, A. and Malcona, J. (1996) Biodiversity and wildlife adaptation to climate change. In: J. Smith, N. Bhatti, B. Jallow, and R. Dixon (eds) *Adapting to climate change: An international perspective.* Springer –verlag, New York, p. 384 – 401.

McAllister, D.; Hamilton; A,; Harvey, B. (1997) Global freshwater biodiversity: striving for the integrity of freshwater ecosystems. *Sea Wind, vol. 11*, p. 1 – 140.

Mc Donald, M.; Hershey, A.; Miller, M. (1996) Global warming impacts in lake trout in Artic lakes *Limnology and Oceanography, vol. 41,* p. 1102 – 1108.

Mdaihli, M.; du Feu, T.; Julius, S.; Ayeni, O. (2003) Fisheries in the Southern border zone of Takamanda Forest Reserve, Cameroon. ST/MAB Seris, WWF, Washington D.C. p. 141 – 154.

Meisner, J. and Shuter, B. (1992) Assessing potential effects of global climate change on tropical freshwater fishes. *Geo Journal, vol. 28*, p. 21 – 27.

Middleton, B. (1999) Wetland restoration, flood pulsing, and disturbance dynamics. Wiley, New York, 388p.

Moss, B. (2000) Biodiversity in freshwater: an issue of species preservation or system functioning. *Environmental Conservation, vol. 27*, p. 1 – 4.

Morin, S. (1982) Notes d'hydrographie et hydrologie camerounaises. Trav. Et doc. Dept. de Geographie, University of Yaounde, 64p.

Ngakfumbe, S. (2001) Rainfall probability and reliability: the case of Cameroon. In: E. B. Eze and C.M. Lambi (eds). *Readings in Geography.* Unique Printers, Bamenda, p. 151 – 175.

Noss, R. (2001) Beyond Kyoto: forest management in a time of rapid climate change. *Conservation Biology, vol. 15*, p. 578 – 590.

Olivry, J. (1978) Transport Solides en suspension au Cameroun. *ONAREST Scientific Papers, vol. 1*, p. 47 – 60.

Poff, N., Allen, J.; Bain, M.; Karr, J.; Richter, B.; Sparks, R.; and Stromberg (1997) The natural flow regime: a new paradigm for riverine conservation and restoration. *Bioscience, vol. 47*, p. 769 – 784.

Pringle, C. (2000) River conservation in tropical versus temperate latitudes. In: P. Boon, B. R. Davis, and G. E. Petts (eds). *Global perspective on river conservation: Science, policy and practice.* John Wiley and Sons, West Sussex.

Reynard, N.; Prudhomme, C. and Crooks, S. (1998) The potential impacts of climate change on the flood characteristics of a large catchment in U. K. *Proceedings of the Second International Conference On Climate and Water: August 1998*, Helsinki, p. 320 – 332.

Rhee, G. and Gotham, I. (1981) The effects of environmental factors on phytoplankton growth: temperature and interactions of temperature with nutrient limitation. *Limnology and Oceanography, vol. 26,* p. 635 – 648.

Saelthun, N.; Aittoniemi, P.; Bergstrom, S.; Einason, K.; and Aamodt, K. (1998) Climate change impacts on runoff and hydropower in Nordic Countries. TemaNord, 170p.

Schindler, D. (2001) Feeding, assimilation and respiration rates of *Daphnia magna* under various environmental conditions and their relation to production estimates. *Journal of Animal Ecology, vol. 37,* p. 369 – 385.

Stemberger, R.; Herlihy, A.; Kugler, D.; and Paulsen, S. (1996) Climate forcing on zooplankton richness in lakes of the northeastern United States. *Limnology and Oceanography, vol. 41,* p. 1093 – 1101.

Ssentongo, G. (1979) *Freshwater fisheries in Africa. Sixth FAO/SIDA Workshop on Aquatic pollution in Relation to the Protection of Living Resources, Nairobi and Mombassa*: 12 June – 22 July 1978. p. 31 – 38.

Teugels, G. and King, R. (1992) Fishes of the Cross-River basin (Cameroon – Nigeria) taxanomy, zoogeography, ecology and conservation. *Mussée de L'Afrique Centrale, Tervuren, Belgique, Annales Sciences Zoologiques 266.*

Chapter Eight

Climate Change And Wetland Ecosystem Management

Summary

Wetlands are geographically isolated ecological niches with a rich biodiversity and a biogenetic potential that can be developed for human welfare. Wetlands also provide important ecosystem services. Unfortunately, these systems are considered as freaks and so have suffered from abuse resulting in stressed conditions. Such stresses reduce their resistance and resilience to future environmental changes. The chapter reviews the current wetland resource potentials of Cameroon, the current anthtropic threats and analyses the observed and anticipated climate change impacts on the system. It concludes that the overlay of human and climate-related stresses will require the selection of management strategies that can enhance resistance to future changes. Consequently the paper presents a framework for designing and managing wetlands to withstand global warming.

Key Words: Wetlands, Climate change, resilience, resistance, conservation, restoration, physical stresses, biological stresses.

Introduction

Wetlands are perpetually or periodically flooded lands. They fall into two categories (Chiras, 1991):

- *Inland wetlands:* These are formed along fresh water, streams, lakes, rivers and ponds. Included in this group are bogs, marshes, swamps and river flow lands that are wet at least part of the year.
- *Coastal wetlands:* These are wet or flooded regions along the coastline, including mangrove, swamp salt, marshes, bays and lagoons.

The formation processes and characteristics of wetlands are controlled by water (Richardson, 1994). These are usually surrounded by uplands; so they have surface water outlet within geographic depressions or transitional zones between terrestrial and aquatic systems where the water table is usually at or near the surface or the land is covered by shallow water. This natural setting therefore determines the nature of pedogenetic process, the development of floristic communities as a function of drainage conditions and both soils and aquatic fauna (Cowardin, 1997). On a global scale wetlands occupy a total land surface area of 8.5km^2 x 10^3 or 6% of the land surface (Maltby and Turner, 1983). A main characteristic is their periodic flooding (Tiner, 1984) and a landscape where precipitation and run-on exceed water losses by seepage and evapotranspiration resulting in impeded drainage (Novitzki, 1979).

An enormous amount of plants, animals and micro organism depend on wetlands. Within their isolated geographical settings the surrounding landscape continuously channels precipitation through the inter-connected lakes, rivers and wetlands. They can therefore be very sensitive to anthropic activities such as extensive land alterations through drainage and water withdrawals. Long-term protection depends on identifying the underlying physical processes of wetlands that are most vulnerable to change, and determining how changes in physical features might affect the resident flora and fauna. Because fresh water ecosystems are sensitive not only to water temperature, volume and flow but also to the variability of the above physical factors wetlands are expected to display a wide variety of changes in response to climate change. The purpose of this chapter is to provide information and strategies for planning and managing reserves that will allow wetlands to withstand and/or adapt to climate change. It reviews the current wetland resource potentials of Cameroon, the current threats they face and the observed and predicted effects of global climate change on them. Finally general suggestions are presented for designing strategies and managing wetlands to increase resilience and resistance to climate change.

Geographical Distribution of Wetlands

Wetlands in Cameroon are geographically isolated ecosystems. They form within large biomes such as the rainforest and the savannah biomes under different topographical and drainage conditions. The main wetland types are discussed in this section.

Mangrove swamps: Mangroves are trees or bushes growing in the subtropical shores. The forest in Cameroon is composed of some 60 species. The dominant species constituting more than 80% of the trees are:

- *Rhizophora racemosa:* commonly known as the red mangrove often attains a height of 25m. The average height falls between 7m and 9m; and
- *Avicennia germinans:* commonly known as the white mangrove often attain a height of 40m. It grows on the more elevated sedimentary parts of deltas that experience periodic floods.

Rhizophora racemosa grows in association with rare tree species such as *Nypa fracticans* and *Guibourtia demensei.* The main fern is *Acrostidum aureum.* Thorn bushes include *Drepanocarpus* and *pandanus parvicentralis nains.* Secondary formations include *Hibiscus tiliaceus, Guibourtia demensei, Drepanocarpus lunatus , Pandanus candelabrum and Phoenix reclinata.* Zones liable to daily flooding are colonized by *Nypa fructicans* while the raised banks of creek meanders are occupied by *Hibiscus tiliaceus, Achrostichum aureum* and *Drepanocarpus lunatus* among others. Zones of transitions between continental formations and swamps are colonized by *Avicennia nitida.* Out of the inter-tidal formations and swamps are the rarely flooded zones colonized by a low forest (4-8m) composed of trees such as *Carapa procera , Raphia nitida, Cynometra mannii, Dalbergia spp.* and *Drepanocarpus lunatus.* Loamy sands and sandy loam in peaty areas support *Rhizophora harisonii, Pandanus candelabrum, Conocarpus erectus* and a herbaceous undergrowth. The creek banks with saline soils are dominated by *Rhizophora harisonii* while the fresh water reaches of creeks are dominated by *Phoenix reclinata* and *Raphia palma pinus* (Morin and Kuete, 1989).

The animal life can be considered under two main ecological niches, that is, mangrove swamps and inshore areas. The aquatic species directly dependent on mangrove includes *Tympanotonus radula, Tympanotonus fuscatus, Pachymelania balteata* and *Neritina glabrata.* Within the stilt-roots of mangroves at the margin of swamps are shellfish, oysters and gastropoda: *Thais haemastoma, Thais Callifera, Littorina scabra, Littorina anguilifera, Melampus Liberians Adams, Cyrenodiarosea, Ostrea tulipa* and *Congeria africana* among others (Plaziat, 1973). See table 1.

Table 1: *The density of fauna dependent on the mangrove swamps.*

Sample site	Species	No. species/m²
Sadi Esengue	*Tympanotonus fuscatus* *Pachymelania fusca* *Neritina*	160 100 2
Defunct ferry crossing at ferry crossing at Dibamba	*Tympanotonus fuscatus* *Neritina*	1024 8
Olga islavd opposite "Bois des Singes"	*Tympanotonus fusca* *Pachymelania fusca Pashymelania aurita*	5 420 10
Pointe Docteur	*Tympanotonus fuscatus* *Pachymelania fusca*	470 260

Source : *Plaziat, 1973*

Each site is dominated by one or two species. This probably demonstrates tolerance to variation in salinity. Thus, *Pachymelamia, Tympanotonus* and *Neritina* demonstrate a high tolerance to changing environmental conditions. These species support the traditional fishing craft in Tiko, Bimbia and Mungo creeks. In Mouanko the most harvested species are *Acrasenelis* and *Tagelus angulatus.* The principal species identified in inshore zones include: *Pseudolisthus elongates, Arius heudeloti, Drepane africana, Pentenemus quniquarius, Pseudotolithus typus, Parapenaeopsis atlantica* and *Palaemon hastaus.*

Mangrove swamps in Cameroon colonize the mouth of the Mungo River, Sanaga and Nyong River mouths and the lower reaches of the Ntem and Londji. These swamp forests are pantropical. They render several economic and ecosystem services. The impacts of climate change on the mangrove swamps have been discussed elsewhere in this book.

Raphia Swamps: Raphia swamps are found along river banks in the zones to the hinterland of the Cameroon Atlantic coast. The zones upstream or to the hinterland of mangrove swamps are colonized by *Raphia vinifera.* Further inland are seasonally flooded prairies or grasslands dominated by *Raphia hookeri* and *Cyrtosperma senegalensis* in Yabassi area.

The *Cesalpiniaceae* forest in the Cameroon Gabon border area has broad river valleys colonized by *Raphia monubuttorum.* The stems are used for the construction of dwellings, the leaves woven as mats for thatching dwellings and the sap extracted as wine. In the forest zones in the Cameroon-Nigeria border these swamps are dominated by *Raphia hookeri* whereas in the Dja forest there is the association with *Raphia monubuttorum* and *Uapaca paludosa.*

Seasonally flooded swamp forests: The rainforest biome possesses numerous seasonally flooded marshy areas. These are riverine areas liable to seasonal floods characterized by trees with elongated roots and a very sparse or patchy grass cover. The trees are dominated by *Uapaca spp.* While the grass cover is ephemeral. A classical example is located in the Nyong River Valley upstream of Ayos. It covers an approximate land surface area of 100.000 hectares along the Nyong and its tributaries. The dominant tree species is *Sterculia subviolaceae.* The soils are humic gley soils that are submerged by a 3 metre depth of water during the rainy season. Between Ayos and Akonolinga these wetlands are characterized by aquatic prairies which probably are secondary successions after the destruction of *Sterculia Subirolaceae* as measure to combat sleeping sickness in the region. The Sangha River seasonally flooded forests are found upstream of Ouesso. Apart from common riverine plant species, these are dominated by *Guibourtia demeusei* and *Irvingia Smithii.* On the banks of the Sangha River are water hycinths such as *Eichornia crassipes.* In Cameroon these are uniquely found only in this valleys whereas they are common in Congo and the Democratic Republic of Congo.

Swamp grasslands: Geographically isolated swamp prairies are found within the rainforest biome of Cameroon. These are found downstream of Ayos and along the Nyong River and the Mbalmayo area. In the Dja forest area these are characterized by seasonal floods. At the confluence of the Mbam and the Sanaga these swamp prairies are surrounded by palms (*Borassus aethiopum*). In the Batouri area they are dominated by *Cyclosorus striatus, Mariscus pseudopilosus* and *Stipularia Africana.* They are also found in Yokadouma to the south east of Djoum.

Gallery swamp forests: Gallery swamp forests are commonly found in the valleys of the Bamenda Highlands and the Bamileke plateau. These are dominated by the *Raphia farinifera.* They are common in swamp forest alluviums of flood plains and river banks where swamp conditions appear to prevail throughout the year, that is, the soils remain waterlogged. *Raphia spp.* grow in association with other trees (Figure 1), shrubs, ferns, lianes, grasses and sedges.

At these locations mounds of micro-relief accumulate around the breathing roots of larger trees, raphia palms and date palms. During the rainy season, pools of stagnant water may accumulate in the depressions in micro-relief. Most of this water drains during the dry season, but the soil remains waterlogged. Finely sorted alluvial soils seem to be ideal. In locations investigated, the soils do not have an organic horizon in the profile. It is mottled right to the surface. The texture increases quickly from a silty clay loam at the surface to a very sticky and plastic clay at greater depths. The profile is gleyed and the permeability is restricted. The gallery swamp forest presents three distinct strata.

- Strata A: several climbing plants and lianes; *Terminalia sp.* and *Mitragyna stipulosa* with heights ranging from 2 to 5 m.
- Strata B: This is the intermediate strata. It is composed mainly of dense stands of raphia palms.
- Strata C: Evenly distributed tall trees ranging from 20 to 30 m in height (See Figure 1).

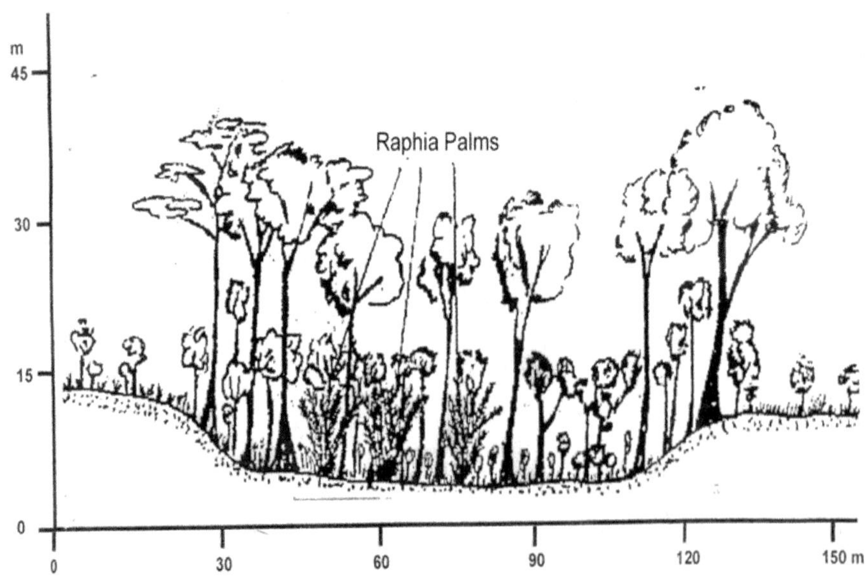

Figure 1: *Cross section of gallery swamp forest in Ndop Plain.*

Ndop Plain Swamp forests and sedge-grass swamps: These wetlands are extensive in the Upper Nun River Valley where they are today under anthropic stress. The construction of the Bamendjin dam in 1975 inundated most of these swamp ecosystems. The following swamp niches can be identified:

- Sedge-grass swamps: These are found in flat areas of the plain where water spreads over the surface. Different sedge-grass communities reflect differences in drainage conditions. The various communities are composed of: s = Sedge, g = grass, f = fern, h = herbs, and t = tree. These communities were largely inundated by the reservoir.

- Water lily swamp communities: found in open water bodies in the plain. *Nyphaea maculate*, and *Nyphaea species* are abundant while *Ultricularia gibba* is frequent.

- Short sedge swamp communities: *Cyperus lanceus* (s) is abundant, *Desmondina salicifolium* (f), and *Scleria melanomphala* (s) are common while *Digitaria abyssinica* is frequent. This community is floating. It is 0.3 to 0.6 m tall.

- Tall sedge swamp forest: This floating community is about 1.5 t0 3.5 m tall. Cyperus species (f), and *Digitaria abyssinica* (g) are abundant. *Cyclosorus straitus* (f), *Cyperus latifolius* (s), *Desmodium salicifolium* (h) and *Jussiaea repens* are frequent. *Melanthera scadeus* (h) is common.

- Short grass swamp community: 0.3 to 0.6 m tall and floating. *Cyperus latifolius* (s) is *Digitaria abussinica* (g) is abundant.

- Fern-sedge swamp community: It is 0.9 to 1.3 m and not floating. *Cyclorosus gongyloides* (f) is abundant while *Commelina diffusa* (h), *Cyperus latifolius* (s) and *Polygronum sp.* are frequent. *Digitaria abyssinica* (g) is common.

- Tall grass and sedge swamp community: The community is 1.5 to 3m tall and not floating. *Cyperus laxespicatus* (s) and *Cyperus nudicaulis* (s) are abundant. *Hyperrhenia bracteata* (g) and *Hyperrhenia rufa* (g) are frequent. *Thelypteris paulustris* (f) is also frequent. *Loudetia phragmitoides* (g) is common and *Raphia farinifera* occurs occasionally.

- Organic swamp forest: it borders the finger-like granitic hillocks (higher ground). It occurs in the form of drowned valleys. Raffia sp. are a very conspicuous component. A black peaty loam overlies slightly decomposed organic material. Trees include *Bridelia sternocarpa, cyathea mannianna, Pandamas candelabrum, Newtonia buchanani, polyscias fulva, fiscus sp.* and some creepers and herbs. The community around Bambalang and south Babessi village was flooded and destroyed to a large extent.

Wetland communities with affinities to those of Ndop Plain are found in Mbo Plain. The two plains have an alluvial morphology reminiscent of a relic lake basin resulting from sedimentation and down-grading by a river (See Figure 2).

Marshes of the Logone River flood plain: The Logone river flood waters extend up to 50km in width and to a height of 1 meter. Elevation ranges from 300 to 350 metres (Figure3). The soils are heavy clays that have wide deep cracks during the dry season. The rainfall itself has little influence on plant growth, but the annual 4 month period of flooding has great influence. The potential plant community is characteristic of open grassland and includes *Hyperrhenia rufa, Sporobolus pyramidalis, Echinochloa stegnia, Setaria palidifusca, Rottobellia exaltata, Echinochloa pyramidalis, Oryza barthii, Sorghum aethiopum*, and *sorghum arundinaceum.* The dominant species are *Hyperrhenia rufa* and *Oryza barthii.* These plants are fire resistant, and when mature they are not palatable to cattle. *Sorghum spp.* are common in disturbed areas and in areas of recent silt deposits. The more palatable plants, for example, *Echinochloa pyramidalis* and *Sporobolus pyramidalis,* are in channeled areas and in the areas distant from water. During drought years, when the Logone River does not flood all of the plain, the vegetative growth is drastically reduced.

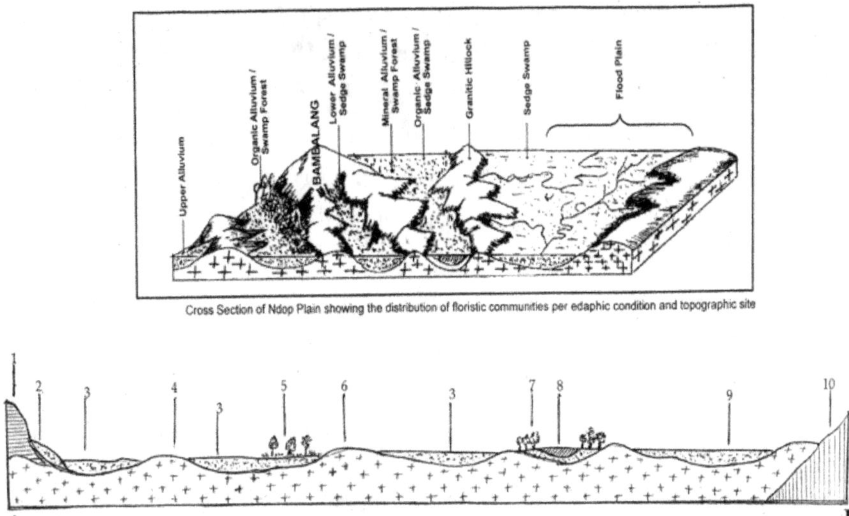

LEGEND: 1. High Lava Plateau, 2: Colluvium, 3: Upper alluvium, 4: Granite/gneiss, 5: Organic alluvium with swamp forest, 6: Bambalang village, 7: Mineral alluvium with swamp forest, 8: Organic alluvium with sedge swamp, 9: Alluviums of the Noun River flood plain, 10: Rhyolite (Mount Mbam).

Figure 2: *Distribution of Ndop Plain wetland communities.*

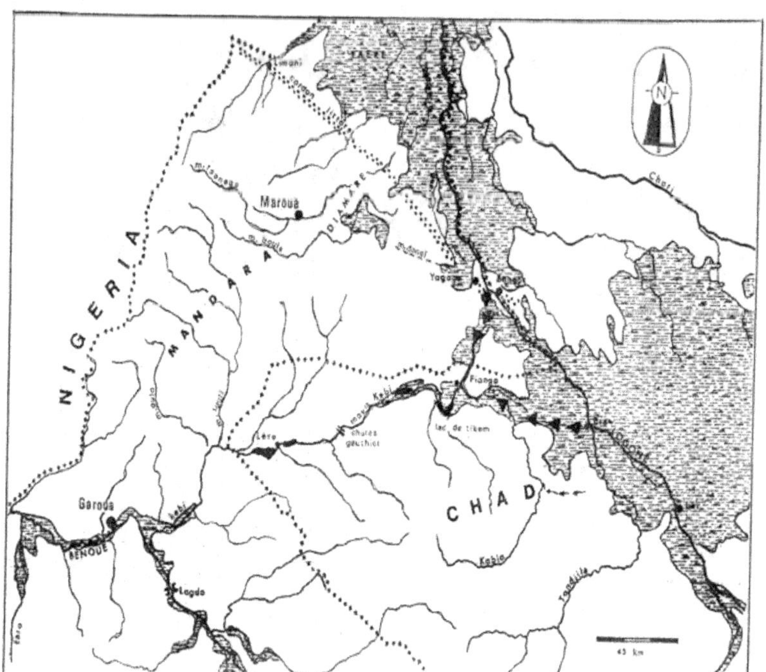

Figure 3: *The Benue-Logone River System showing water spillages from the Logone into the benue Basin via the Mayo Kebi and the Yaeres (marshes)*

These flood plain prairies have one of the highest forage yields in North Cameroon. Production is 8.000kg per hectare in unfavourable years and 10.000kg in favourable years. Wildlife and livestock annually leave the Yaeres for the higher, sandy range sites because of the clayey soils and the flooding that occurs from May through September. This migration provides time for the grasses to grow before the grazing animals return during the dry season. Flooding and repeated burning by hunters and herders have prevented woody vegetation from growing in the Yaeres.

Photo: *Receding flood water in the Grand Yaeres attracts both cattle and wildlife during the dry season.*

The Yaeres have a total land surface area of 141,200 hectares. The soils are flooded for 3 to 5 months of the year but drain very fast when floodwaters reduce. They have a very high potential for dry-season sorghum and rice cultivation. It is under the pressure of drainage for wetland pady, overgrazing and annual burning.

The Value of Wetlands

These wetlands are an extraordinary valuable and productive fish and wildlife habitat. Many other animal and plant species also inhabit these ecological niches, for example, the deer and others live in or around wetlands. In addition, shellfish, amphibians, reptiles, birds and fish also live in these endangered habitats.

Aside from their importance to wildlife, wetlands also play an important role in regulating stream flow. Wetlands act as sponges, holding back rainwater and reducing natural flooding.

Photo: *Extensive wetland riziculture in Doumaraye village on the Logone River flood plain (IGN67-68 CAM 01/200 No. 189).*

They also filter out sediment eroded from the land and therefore help reduce sedimentation of streams. Wetlands also act as traps for nitrogen and phosphorus, two common pollutants washing from heavily fertilized land. Wetland plants absorb these nutrients, preventing water pollution downstream. The sponge effect reduces flooding and has the added benefit of recharging groundwater. Coastal wetlands help protect human settlements by absorbing many storm surges, high waves that accompany high winds. Wetlands are used to grow certain cash crops such as rice (Figure 4). This is the case in the Ndop Plain, Yaeres, Menchum River floodplain and Mbo Plain. Because their usefulness is not always apparent, they are often filled in or dredged to make way for housing, recreation, industry and garbage dumps. In the West Cameroon Highlands these ecological niches have been drained for vegetable cultivation.

Current Threats to Wetlands

Physical barriers to flow:

Physical barriers on rivers such as dams to generate power, and to supply water for irrigation alter the natural flow of water and therefore can adversely affect wetland ecosystems. Initial flooding of the Upper Nun River valley following the construction of the Bamendjin dam to create a reservoir resulted in massive losses of terrestrial and wetland habitats in Ndop plain. Dams also cause dramatic changes in downstream flow regime, where seasonal, dynamic flows are replaced by steady water releases for energy production, or intermittent large releases to lower reservoir levels. The loss of natural silt in released water can lead to a range of damaging effects downstream, including changes in chemical composition, river bank erosion, and massive habitat loss and erosion in coastal deltas and floodplain wetlands. These barriers disrupt connections between rivers and valuable floodplain wetland habitats, which serve as refugia and spawning grounds for many animals, and are often sites of high biodiversity.

Water diversions and withdrawals:

In order to meet the agricultural needs of a growing population, large quantities of water are withdrawn from rivers or the underlying water tables. Such withdrawals and diversions are already affecting the Yaeres in the Logone River floodplain. This is resulting in reduced seasonal floods necessary for maintaining wetlands. The development of large-scale irrigation schemes such as in Ndop Plain and Yagoua in the Yaeres has considerably reduced the size of wetlands (Figure 4). Reduced seasonal flows and flooding, variability in flows all lead to habitat loss. The withdrawal of groundwater for agricultural purposes in adjacent areas can completely dry up valuable wetlands, even if habitat itself is protected (McDowell, 1984).

The large scale degradation of gallery swamp forests in the Bamenda and Bamileke highlands has been reported by several researchers (Ndenecho, 2007, Kometa, 2009, Elang *et al*,, 2007 and Dongmo, 1986). The drainage and conversion of gallery swamp forests in the region is threatening the future existence of raphia palms.

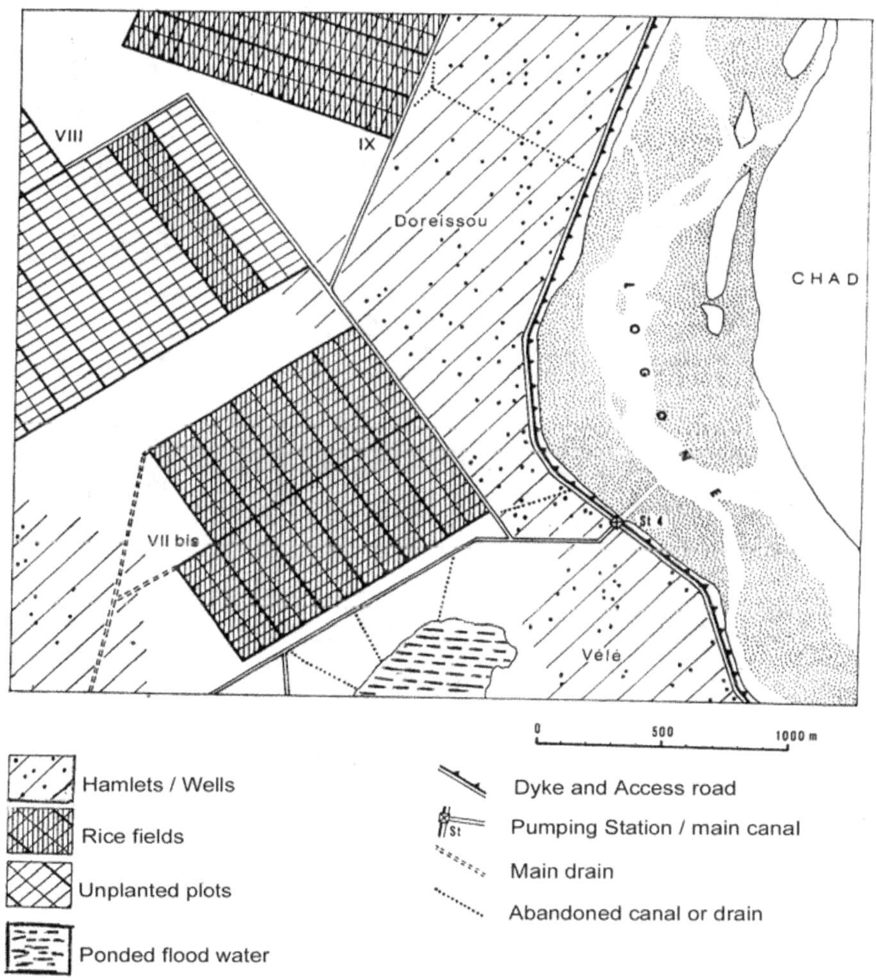

Figure 4: *Layout of irrigated rice fields in Doreissou village (Source: National Geographic Institute aerial photograph IGN 78 CAM 131/200 No. 54) on the Logone River floodplain.*

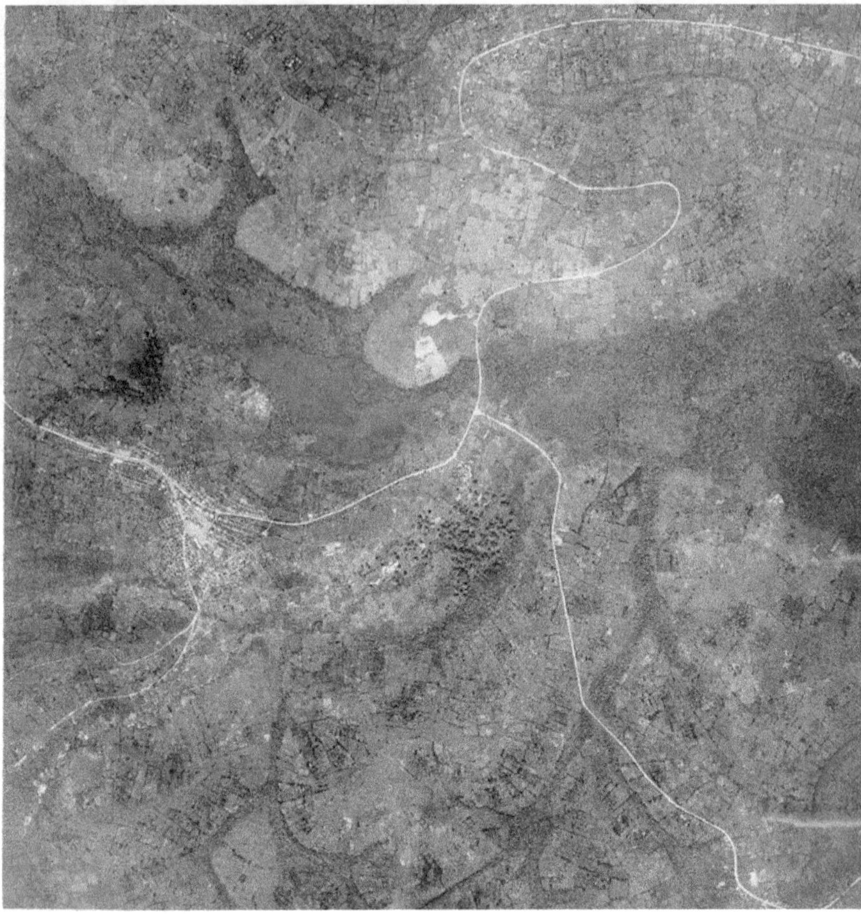

Photo: *Aerial photograph of the Bamendjou-Balessing area showing the distribution of gallery swamp forest (IGN phot NB-32-X1-1C 200)*

Figure 5: *The distribution of gallery swamp forest in Bamendjou (West Region): A=gallery swamp forest, B= Life fence enclosures, C= village service centre. Gallery swamp forests are found in low-lying valley bottoms. These are being converted to vegetable plots.*

Table 2: *Distribution of land in reclaimed gallery forests in some villages of West Region.*

Farm size (m2)	Gallery swamp forests		
	Batsingla	**Fokamezo**	**Balefok**
Less than 400	5	5	0
400 - 599	50	53	2
600 – 999	24	41	4
1000 – 1999	17	47	72
2000 – 4999	10	9	15
3000 – 10.000	0	0	1
Over 10.000	0	1	0
Farms involved	161	339	712

Source: Dongmo, 1986

Table 2 presents some data on the systematic drainage of swamp forest in the Bamendjou area of the Bamileke plateau. This has resulted in a spectacular transformation of the landscape with negative impact on the survival of the raphia palm. The degradation and conversion of natural raphia palm forest to farmlands and market gardens in the highlands is widespread (Ndenecho, 2006; Ndenecho 2007). If this is allowed to continue, not only will the raphia palm be eliminated in the region but also the products and livelihood activities it supports. Craftwork dependent on the raphia palm is the traditional and cultural heritage of the indigenous people. It deserves to be preserved and promoted, for it gives life, colour and pride to the community. It is the cement that binds the numerous ethnic groups together and by it, even the post- modern world; they may regain self respect and begin to re-discover their cultural identity (Knöpfli, 2001).

Filling and drainage of wetlands

Wetlands throughout the country suffer from filling and draining for agriculture and housing developments. In urban areas wetlands have been filled and drained to the extent that their natural role in flood and storm water mitigation has been lost. Flood disasters associated with filling and drainage of wetlands are increasing in frequency and intensity in cities such as Limbe, Douala, Yaoundé and Bamenda. These habitats have traditionally been seen as useless in their natural state. The state has in most cases provided funding to convert wetlands to agricultural land such as Ndop plain, the riziculture project in Yagoua, riziculture project in Mbo plain and the Baigom Agricultural Development Project (Figure 6). Today, the valuable ecosystem services that wetlands provide, including water purification, groundwater recharge, and flood control, are gradually being appreciated by government agencies and municipal managers. Drainage can lead to changes in species composition and distribution.

Pollution of wetlands

Large quantities of pollution enter wetlands from point services such as industrial and municipal sewage outflows. In most urban areas both solid and liquid wastes are dumped in wetlands and riparian zones. Runoff from fertilizers used in commercial agriculture add large amount of nitrogen and phosphorus to wetlands. Such nutrient enrichments result in the excess growth of algae which is sometimes toxic to wetland life because it results in reduced water clarity and light penetration. Because of this increase primary productivity, the activity of decomposition, oxygen-consuming bacteria increases and oxygen level decline. Shifts in the food web and alterations in wetlands habitat can lead to changes in species composition and distribution. For example, the density, distribution and relative abundance of wetland plants can change after eutrophication (Schmieder, 1997), and valuable food species are often replaced by less desirable fauna that can tolerate low oxygen levels (Egerton, 1987).

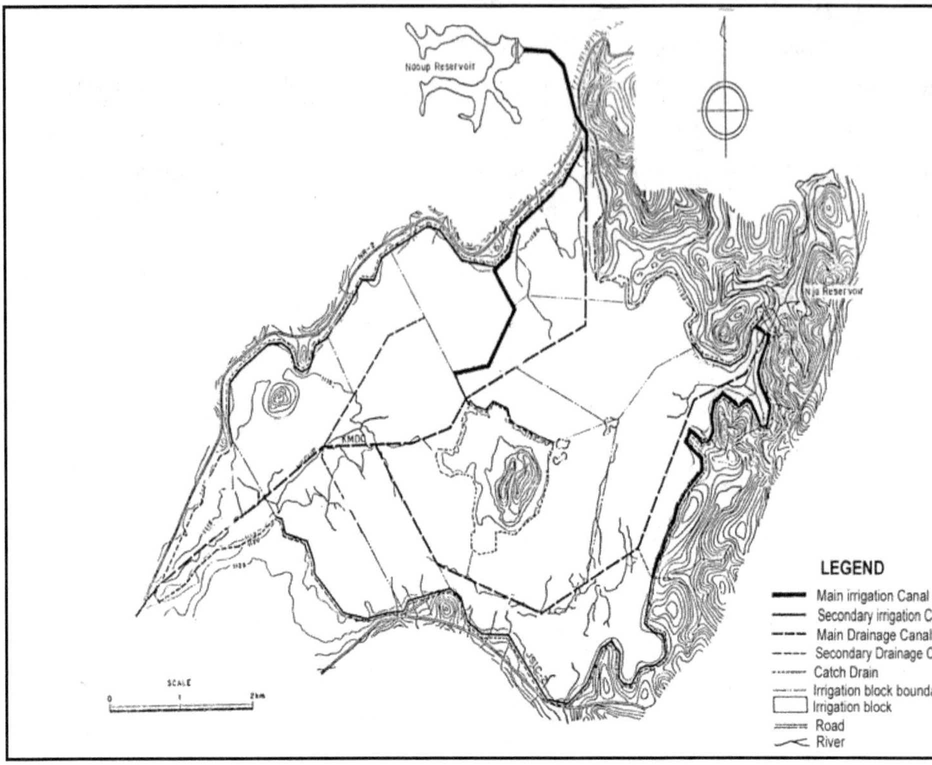

Figure 6: *Irrigation and drainage project in Baigom wetland funded by government of Cameroon and Japan International Cooperation Agency.*

Potential Wetland Biogenetic Resources

The menace of water weeds in some parts of the world is reaching alarming proportions. Water is an important resource but aquatic weeds affect it adversely by blocking canals and pumps in irrigation projects; interfering with hydroelectricity production, wasting water in evapotranspiration; hindering boat traffic, increasing water-born diseases, interfering with fishing and fish culture; and clogging rivers and canals so that drainage is impossible and floods result.

As aquatic weeds spread, they disperse that water snail that causes *schistosomiasis,* the insidious debilitating disease prevalent in the wetland areas. The snails live in the dangling roots of floating plants and ride around as wind and current float the plants around. In addition these plants foster malaria.

And yet, in a sense, aquatic weeds constitute a free crop of great potential value-a highly productive crop that requires no tillage, fertilizer, seed or cultivation. These plants according to the National Academy of Sciences (1976) have potential for exploitation as animal feed, human food, soil additives, fuel production (biogas) and waste water treatment. As synthetic materials for pulp, paper and fibre become increasingly expensive in the future these aquatic materials may be used for these purposes. *Phragmites communis* is one of such weeds. *Cyperus papurus antiquorum* also have this promising potential.

In order to fully exploit the potentials of wetland resources immense research is required. Such research efforts should go along with policy to protect these geographically isolated ecological riches that are found all over the country.

Pysical Effects of Climate Change on Wetlands

Increased air temperatures are likely to have a drying effect on wetlands, unless increased rainfall compensates for the evaporation. Shallow and ephemeral habitats such as depressional wetlands with no channelized flow in or out of wetlands in the direr regions of North Cameroon could be lost entirely, especially if rainfall declines and man is obliged to intensify the extraction of ground water reserves in the logone (Gitay *et al*; 2001). On the whole a drier climate resulting from global warming is likely to lead to the shrinkage and loss of wetland habitats, as will increase habitat fragmentation.

Conversely increase precipitation resulting from extreme rain storm events could lead to flooding, expansion and deepening of wetland habitats, and increased connectivity in riverine systems. However, increased rainfall or extreme flooding may also lead to increased input of sediments and woody/plant debris and pollutants. This could destroy some wetlands if vegetation or important habitat features are completely submerged. Coastal wetlands could be destroyed in this way.

Coastal freshwater wetlands will be particularly sensitive to high tides resulting from an increase in storm frequency or magnitude. These high tides can carry salts inland to salt-intolerant vegetation and soils, and could lead to the displacement of freshwater flora and fauna by salt-tolerant species (Michener *et al.,* 1997). Rising sea level could destroy coastal freshwater wetlands as saline water invades (Tyedmers and Ward, 2001).

Biological Effects of Climate Change on Wetlands

Ephemeral, depressional wetlands, especially those in arid and semi-arid areas, often harbour rare species that would be lost if these areas dry up (Belk and Fugate, 2000). The already threatened habitat loss in the Yaeres could disappear if reduced rainfall and increased evaporation eliminate the shallow, vernal habitats.

Small temporary wetlands occur in all the biogeographic landscapes of Cameroon. Such wetlands are often used by more species than permanent ponds (Gibbs, 1993). The drying and loss of wetlands would reduce not only the number and the size of available ponds, but also increase inter-pond distance (Gibbs, 1993); Semlitsch and Bodie, 1998), lowering chances of amphibian recolonisation, since adult frogs are generally only capable of travelling 200 to 300 metres (Sjogren, 1991); Skelly *et al.;* 1999). Drying and loss of wetlands would also reduce habitat connectivity on regional scale, endangering migrating birds that depend upon a network of wetlands along their migration route (National Research Council, 1995).

Management of Wetlands to Withstand Climate Change

Wetland like mangrove swamps have generally been considered as freaks. They have therefore received little protection attention. As a first step towards protection, there is a need to make an inventory and classification of wetlands in terms of their biogeographic, ecologic characteristics, and geomorphic processes within their physical settings. This is important because high biodiversity is often found in these isolated and heterogenic habitats – especially dynamic habitats with seasonal changes in water level (Abell *et al,* 2002). Planning and management guidelines for these wetlands include:

Protection of physical features around wetlands

The ecosystem function in wetlands is determined by physical characteristics such as swamp morphology, water flow volume, drainage status, seasonal variation in water level, sediment input, nutrient balance, and water quality (Abell *et al;* 2002). Conservation efforts must therefore focus on the protection of these physical characteristics. The physical features of wetlands are expected to undergo a number of changes as a result of climate warming and rainfall variability.

Preserve habitat connectivity

The preservation of habitat connectivity is necessary to enable access to migration routes. Connectivity is important not only between different freshwater habitats that are between rivers, lakes and wetlands, but also along the length of a river. In the face of climate change, connectivity may allow fauna to migrate to more suitable habitats or thermal refugia.

Reduction of human pressures on wetlands

The human-related threats to wetlands such as filling, drainage, pollution, conversion to other land uses and the introduction of exotic species need to be controlled. Such stresses reduce the resistance and resilience of wetland flora and fauna to climate warming. Many wetlands will also be more liable to invasions as drainage conditions change as a result of reduced precipitation and increased temperatures. The control of exotic or invasive species will therefore become an important aspect of management.

Regulate extractive processes within wetlands

Wetlands can be extremely sensitive to distant, upstream disturbances, such as agriculture, deforestation, diversions and other human processes that change the physical characteristics of the habitats. Wetland ecosystem function will be affected by such extractive processes as drainage, dykes, diversions, reclamation, irrigation agriculture, nutrient enrichment and over exploitation of biodiversity. These ecosystems are intricately connected to one another and to the terrestrial systems that surround them. Common reserve considerations such as the reserve size and the creation of buffer zones around them are applicable. Habit alteration resulting in stresses will reduce the resistance and resilience of these systems to climate change.

Restoration of degraded wetland sites and response to climate–related threats

Degraded wetland ecosystems normally suffer from biological stress. A stressed ecosystem exhibits reduced resistance and resilience to climate change. The restoration of degraded wetlands will therefore enhance resistance and resilience to future change. It will also enhance the recovery of the degraded wetland. Successful restoration techniques include restoring the hydrology of wetlands by removing impediment to flow (Gilbert and Anderson, 1998), elimination of human-related stresses and the restoration of the physical site, flow and drainage conditions. In urban areas such wetlands can be converted into aesthetic ponds by environmental biologists and landscape planners. In this way the ecological integrity of wetlands can be improved. The transplantation of fauna and flora within isolated wetlands in a biogeographic region may be envisaged. It should, however, be emphasized that such transplanting can disrupt the ecosystem function of the existing local wetland system. This will therefore constitute a bad idea. Transplanting vulnerable wetland species to new, created habitats, where no native community exists (uninhabited, enclosed, newly created wetlands) would eliminate the risk of damaging an existing biotic community.

Conclusion

There are both climate-change and human related threats to the survival of wetland ecosystems. The application of a general list of conservation goals blindly ignores the specific strengths and weaknesses of the system, and does not account for the relative importance of various threats. There is need to undertake site-specific evaluation of all factors and strategies to reconcile conflicting goals in a wetland area. This should constitute the basis for designing a successful wetland adaptation strategy. The selection and implementation of adaptation strategies in a particular isolated geographic area should consider the following:

- characterize and monitor species that are most vulnerable to climate change in order to determine the scale of conservation needs;
- characterize and monitor wetland ecological niches and systems to determine the scale of conservation needs;

- develop strategies for protecting and restoring wetland ecosystems while dealing with increasing human-related stresses, and

- perform integrated river basin management (IRBM) and ecoregion conservation (ERC) in order to:

(a) buffer systems against climate change and provide a basis for long long-term conservation of inter-related freshwater systems; and

(b) protect entire freshwater ecoregions to preserve the ecological and evolutionary driving forces of biodiversity.

References

Abell, R.; M. thieme; Dinerstein, E; Olson, d. (2001) A source book for conducting biological experiments and developing biodiversity visions for ecoregion conservation. Volume II: Freshwater Ecoregions. WWF, Washington, D.C. 2001p.

Belk, D. and Fugate, M. (2000) Two new Branchinecta (*Crustacea, Anostraca*) from the South Western United States. *The Southwestern Naturalist Vol. 45,* p. 111-117.

Chiras, D. (1991) Environmental Science. The Benjamin/Cummings Publishing company, INC. 549p.

Cowardin, L., Carter, V., Golet, F., Laroe, E. (1997) Classification of wetlands and deep water habitats of the United states. U.S. Department of Interior, Fish and Wildlife Service. Odffice of Biological Service. Washington, D.C.

Dongmo, T-L. (1986) Amenagement et mise en valeur des grands bas fonds aux sols hydromorphes en pays Bamelike. *Cahier Geographique*, Department of Geography, University of Yaounde.

Engerton, N. (1987) Pollution and aquatic life in Lake Erie: early scientific studies. *Environmental Review, vol. 11,* p. 189-205.

Elong, J.; Kamga, A.; Bounou, V. (2007) Les nouveaux enjeux socio-économiques de la mise en valeur agricole des bas-fonds à Penka-Michel dans la Menoua. *Annals of the Faculty of Arts, Letters and Social Sciences, Vol. 1,* p. 29-53.

Gibbs, J.P. (1993) Importance of small wetlands for the persistence of local populations of wetland associated animals. *Wetlands, Vol. 13*, p. 25-31.

Gilbert, O.; Anderson, P. (1998) Habitat creation and repair. Oxford University Press, New York. 288p.

Gitay, H.; Brown, S.; Easterling, W.; Jallow, B. *et al.* (2001) Ecosystems and their goods and services. In: J. McCarthy, O. Canziana, N. Leary, D. Dokken, and K. White (eds). *Climate change 2001: adaptation and vulnerability*. Contribution of working Group II, 3rd Assessment Report of the IPCC Cambridge University Press, Cambridge, p. 235-342.

Knopfli, H. (2000) Basket, and calabashes, palms and people: some traditional craftsmen and craftswomen of the Western Grassfields of Cameroon. Basel Mission Press, Basel. p. 125-142.

Kometa, S. (2009) Wetland exploitation and its environmental impact: the case of the Bamenda-Bafoussam axis of the Western Highlands of Cameroon. *Proceedings of the Postgraduate Seminar on conflict Prevention, Management and Resolution.* Dept. of Geography, University of Buea. p. 25-37.

Maltby, E. and Turner, R. (1983) Wetlands of the world. (1983) Wetlands of the world. *Geographical Magazine,* p. 12-17.

McDowell, R. (1984) Designing reserves for freshwater fish in New Zealand. *Journal of the Royal society of New Zealand, Vol. 14*, p. 17-27.

Michener, W.; Blood, E.; Bildstein, K.; Brinson, M. and Gardner, L. (1979) climate change, hurricanes and tropical storms, and rising sea level in coastal wetlands. *Ecological Applications, Vol. 7*, p. 770-801.

Morin, S. and Kuete, M. (1989) Le littoral camerounaise: probléme morphologiqes. *Trav. De Laboratoire Géographie physique Applique.* Université de Bordeaux III Talence.

National Academy of Sciences (1976) Making aquatic weeds useful: some perspectives for developing countries. N.A.S, Washington D.C. p. 1-10.

National Academy of Sciences (1995) Wetlands: characteristics and boundaries. National Academy Press, Washington D.C. 307p.

Ndenecho, E. ((2005) Savannization of tropical montane cloud forest in Bamenda Highlands, Cameroon. *Journal of the Cameroon Academy of Sciences, Vol. 5,* p. 3-10.

Ndenecho, E. (2007) Population dynamics, Rural livelihoods and forest protection projects in Sub-Saharan Africa: experiences from Santa, Cameroon. *International Journal of Sustainable Development and World Ecology, Vol. 14.* In Press.

Novitzki, P. (1979) Hydrologic characteristics of Wisconsin's wetlands and their influence on flood stream flow. In: P. Greeson, J. Clark (eds). *Wetland functions and values: the state of our understanding.* American Water Resource Association, Mineapolis, p. 377-388.

Plaziat, J. (1992) Repartition de mollsques amphibies de quelques littoraux et estuaire à mangrove (Nouvelle-Calédonie et Cameroun), role de la salanité dans les modifications locales des peuplements mangrove. CECA-CEE-CEEA, Bruxelles.

Richardson, C. (1994) Ecological functions and human value in wetlands: a framework for assessing forestry impacts. *Wetlands, Vol. 14,* p. 1-9.

Schmieder, K. (1997) Littoral zone – GIS of lake Constance: a useful tool in lake monitoring and antecological studies with submersed macrophytes. *Aquatic Botany, Vol. 58*, p. 333-346.

Semlistsch, R. and Boodie, J. (1998) Are small isolated wetlands inexpendable? *Conservation Biology vol. 12*, p. 1129-1133.

Sjogren, P. (1991) Extinction and isolation gradients in metapopulations: the case of the pool frog *Rana lessonae*). *Biological Journal of the Linnaean Society, Vol. 42*, p. 135-147.

Skelly, D.; Warner, E.; Cortwright, S. (1999) Long-term distributional dynamics of a Michigan amphibian assemblage. *Ecology, Vol. 80*, p. 3326 – 2337.

Tiner, R. (1984) Wetlands of the United States: current status and recent trends. United States Fish and wildlife Service, Washington D.C.

Tyedmers, P. and Ward, B. (2001) A review of the impacts of climate change on BC's freshwater fish resources and possible management responses. *Fisheries Centre Research Reports, Vol. 9*, p. 1-12.

Chapter Nine

Climate Change and Management of Physical Barriers in Riverine Systems

Summary

The construction of large dams on riverine systems is a subject of great controversy. Environmental effects have been well documented. The basic point at issue is whether their benefits outweigh their environmental and social costs. This chapter makes an assessment of the human-related physical barriers constructed on riverine systems in Cameroon and identifies the conflicting resource development and management goals which may arise in the face of increasing global warming and the use of dams for development purposes. It reviews the ecological implications of dams and man-made lakes and establishes how dams can conflict with measures to enhance the resistance and resilience of riverine ecosystems to climate change. The chapter finally concludes that there is a needed change in paradigm to the use of a series of smaller dams on rivers. These could provide most of the benefits at lower investment cost, greatly reduce social costs and environmental costs, and enhance the adaptations of riverine ecosystems to global warming.

Key Words: Dams, man-made lakes, ecological effects, ecosystem connectivity, aquatic life, aquatic habitats, sustainable management.

Introduction

Far back into history, man has undertaken the modification of rivers mainly to ensure ready and reliable availability of water. Rivers and streams have been dammed to store water, sometimes for single use but increasingly for multi-purpose development projects. It has been estimated that there are about 10,000 artificial lakes with surface area exceeding 1,000 km^2, scattered round the world. In addition, millions of small rivers and streams have been dammed and there are so many small impoundments that it is almost impossible to gauge their numbers (Obeng, 1979).

In Africa these dams play an important role in rural water supply for domestic and agricultural use. There are more than 20,000 small dams in the Central African Republic, about 12,000 in the Nyanza Province of Kenya, and many in Northern Ghana, Sudan, Nigeria and Uganda. Such small impoundments are also a common feature of China, and in India almost all major and minor rivers have dams on them. Cameroon has also constructed several dams.

The chapter assesses the human-related physical barriers and alterations to hydrology as a basis for resolving development and resource conservation conflicts with increasing human water demand and global warming. It reviews the environmental implications of dams and man-made lakes as opposed to natural water bodies and identifies the scope for building the resistance and resilience of aquatic systems in the management of these structures with increasing global warming.

Dams and Reservoirs

Hydro-electric power in Cameroon is obtained from water. The largest power station is built at Edea on the Sanaga River fall where a dam 1600 metres high has been built to hold back water. It is the third largest dam in West Africa, the first two being Kainji on the Niger River in Nigeria and Akosombo on the Volta River in Ghana. These two stations each generate 750,000 kilowatts, while Edea between 150,000 to 250,000 kilowatts (Ngwa, 1979).

To generate electricity, a dam is built across a river on top of the fall. Below the fall, a power house in which a large turbine and electric generators are installed. Water is directed through huge steel pipes from the reservoir above, behind the dam down into the power house, where it falls into the blades of the turbine, causing them to rotate. This turns the electric generators and so produces electricity (Figure 1). The speed at which the turbine revolves is dependent on the height of the fall, and the amount of water in the fall and the amount of water in the reservoir. This means that the level of water in the reservoir must be kept high and constant so that there is no drop in the amount of electricity produced. It is for this reason that two additional dams – the Mbakaou and the Bamendjin dams were built on the Djerem and Nun Rivers, to increase the water in the Edea station during the dry season. The capacity of the Mbakaou dam is 2.9 milliard cubic metres of water while that of the Bamendjin is 1.8 milliard cubic metres of water (Ngwa, 1979). See Figure 1.

Figure 2 presents the Bamendjin dam. It is 245 m long and 17 m high and is built on the upper course of the Nun River, a tributary of the Mbam River which flows into the Sanaga River. The lake behind the dam forms an artificial lake that is 344 km^2. On the other hand, the Mbakaou dam is on the Djerem River, a tributary of the Sanaga River, situated south east of Tibati. The aim of these dams is to increase the discharge of the Sanaga River, and consequently increase the power and alluminium production at Edea. Associated with the Bamendjin reservoir is an extensive riziculture based on irrigation and enterprising craft fishing.

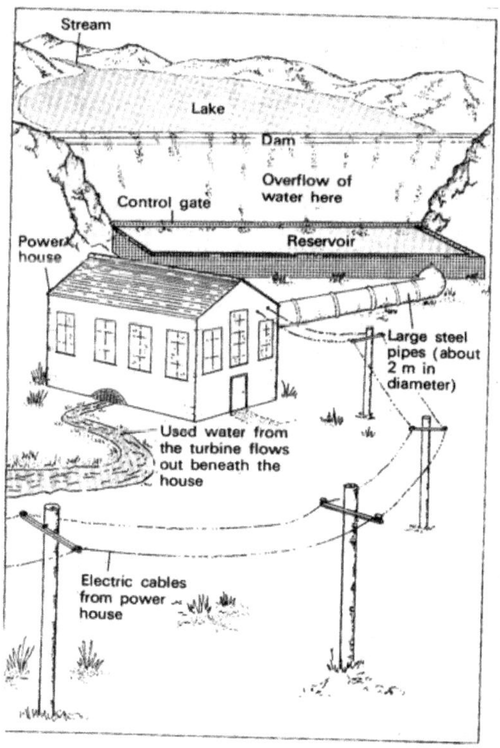

Figure 1: *Sketch of the Malake (Meanja) defunct hydroelectric power station in South West Region on the Yoke River (Ngwa, 1979).*

Photo: *Edea hydroelectricity station and dam*

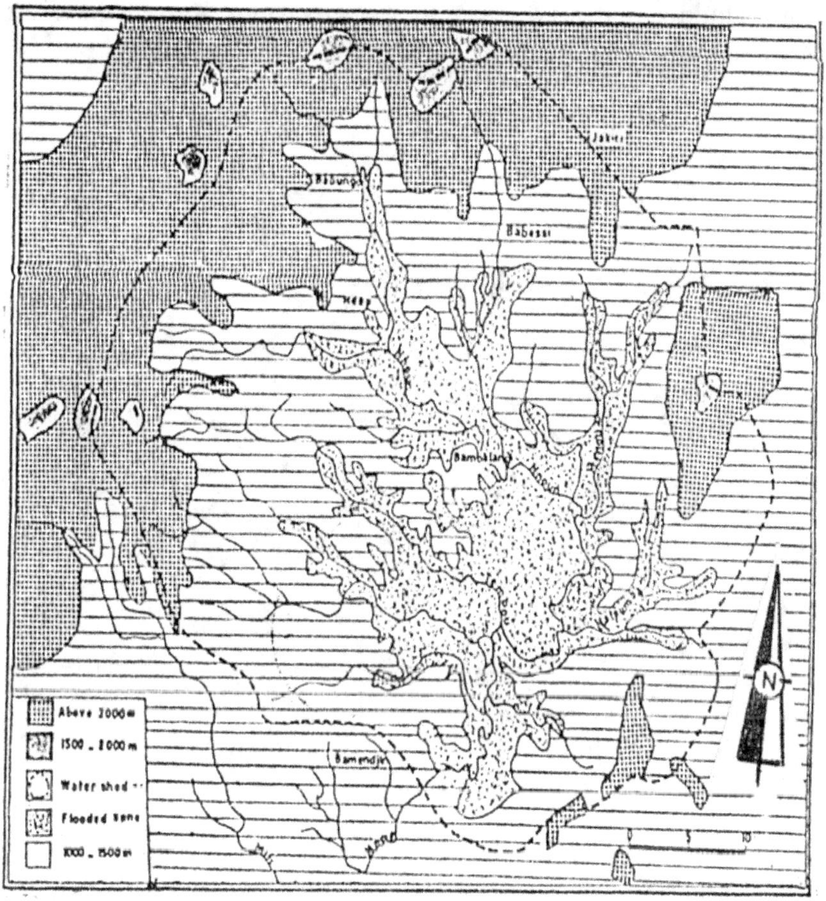

Figure 2: *Regional setting of the Bamendjin Reservoir (Upper Nun River Valley)*

Other dams producing hydro power include the Song-Loulou on the Sanaga River, some distance from Edea. It has a capacity of 240 mw, and is linked to the main station at Edea. The Njock-Mpoume is built on the Nyong River and is estimated to produce 232 mw. The Lagdo dam is on the Benue River. It impounds 4 milliard cubic metres of water and produces 72mw of electricity. Minor dams producing hydro power include the Mefou dams on Mefou River and Mallale dams on Yoke River (defunct).

The Chad basin possesses several micro-dams, diversions and reservoirs that store water solely for agricultural purposes. The main diversion and dam sites are in Ere, Toubouro, Lere, Moundou, Mbourau, Geidam and Gashua. Several other small dam projects in this area are envisaged and have been discussed elsewhere in this book. Figure 3 presents the Maga reservoir in the logone river floodplain. Water is diverted by the canal from the logone River for impoundment. It serves an extensive riziculture project in the Yagoua area (Figure 3).

Photo: *The Nyong River at old ferry crossing.*

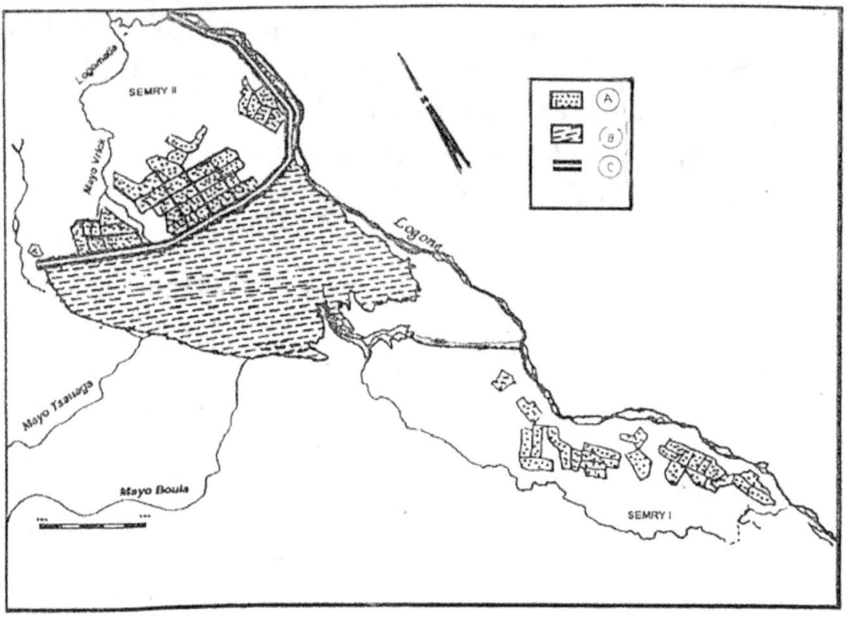

Figure 3: *Location of Maga Reservoir and irrigated zones: A= irrigated zones, B= reservoir, C = dykes (Source: Molinier et al., 2000)*

Extensive floodplain drainage schemes for agricultural development projects abound in the country. These include the Yaeres of North Cameroon, Ndop Plain, Mbo Plain and the Nyong River Basin. Such projects and diversions alter the natural ecological equilibrium of riverine systems.

Natural rivers carry abundant silt each year to the land stretching along their banks and to the sea. In the sea this nutrient-rich silt nourishes a variety of micro-organisms that, in turn are food for thriving fish populations. On the land the silty water floods nutrient – impoverished soils as the rivers spill over their banks, enriching the soil in these areas. The construction of dams in many African countries focused primarily on the benefits that would result from the dams versus the economic cost of construction. Little attention was given to possible ecological backlashes.

For example, not long after the Aswan Dam was constructed in the 1960s and Lake Nasser began to fill, the people of Egypt and the world grew alarmed. Numerous problems began to make themselves painfully evident:

- the periodic flooding that had provided the annual fertilization of the land ceased;
- the sardine fishery in eastern Mediterranean collapsed;

- the rising waters of Lake Nasser threatened the Ramses Temple at Abu Simbel, built over 3000 years ago. Engineers and construction workers dismantled the huge temple piece by piece and moved it to a site 60 metres above its original level, where it would be safe from the rising waters; and
- the incidence of schistosomiasis increased as a result of the dam. The organism that causes this disease is carried by snails. Snails require a constant supply of water, which the lake and the irrigation canals provided.

If a cost-benefit analysis had been done before construction of the dam began, chances are that it might not have been built. With increasing global warming dams and diversions weaken the resistance and resilience of riverine ecosystems.

Characteristics of Natural Lakes and Rivers

Bodies of standing and barely moving water are not new; nature has always had its own lakes. Of the African lakes whose sizes are known, there are over 30 which with areas ranging from 120 to 800km^2. Such lakes can be described as "old" and generally they have stabilized physically and ecologically. When unpolluted and in a healthy state, they support a wide variety of freshwater animal and plant life in a medium with a specified range of physical and limnological characteristics. They have a variety of origins. Those caused by glaciers are usually limited in distribution but others, which result from landslides and other earth movements or where river channelling forms oxbow lakes, are universal in distribution. Natural lakes can have great depths. Lake Baikal, for instance, is estimated to be about 1,706 m deep and Lake Tanganyika, the second largest African lake, is 700 m deep. Depth, surface area and the character of the margin generally influence the ecology of lakes.

Natural lakes have their own characteristics which make them different from artificial impoundments. Foreign material, algal and other remains and deposits are formed in their depths. Chemical analysis have shown that in addition to organic matter, such deposits may also contain, in varying proportions, oxides of iron, (Fe_2O_2), Calcium (CaO), magnesium (MgO), phosphorus (P_2O_5), aluminium (Al_2O_3) and silicon (SiO_2). The water itself carries chemicals in solution. The hydrogen ion concentration of natural waters may give a pH range of 3.2 to 10.5, but this is by no means a general condition and the pH is subject to changes annually, seasonally and even diurnally on small and protected lakes. The concentration and distribution of chemicals, including gases, and the range of temperature in the medium, to a large extent influence the type and abundance of micro-organisms present, while location, topography of the adjacent land and shore line characters contribute to the definition of the biotic features of the lake. The combined physical and chemical nature of a lake therefore largely determines its ecological character. Generally, where there is no drastic outside interference, the biological relationships are reasonably stable. The fauna is adapted to living in slow waters and they are seldom sedentary since they have to find their food. Snails and worms are examples of typical lake animals (Obeng, 1979).

Rivers with their moving environments maintain ecosystems which, while varying in detail, remain characteristically riverine in terms of the faunal diversity and abundance. River fauna tends to be specific to particular niches which suit their feeding and breeding habits. Swift waters with stable rock bottom support a range of animals different from those adapted to slow water with muddy beds. Usually, pelagic communities are few and plankton may be scantier than in static conditions. The invertebrates, which are sedentary, often have effective "holding" mechanism, and filter food from the fast current. The larvae of *Simulium* species, for instance, hold fast to substrates in swift waters and when dislodged they hold on to a fine yarn. They also cement their eggs which are laid in fast water to objects. Other animals have a well developed power of swimming to stay in the strong currents (Obeng, 1979).

Biological and Physical Effects of Damming

The dams in Cameroon have witnessed little or no limnological studies. The biological and physical impacts of damming on the environment are little known. This section therefore reviews the experiences of man-made lakes in this sub-Saharan African Region.

The construction of a dam across a river literally causes an ecological revolution. Riverine physical and chemical conditions and the biological communities that they normally support are destroyed. The obstruction cause a swelling of the river size, a wideways spread of water and, except for the original river-bed, the new water body lacks the great depth of most natural lakes. The resultant water quality is very different to that of the river and to that of shallow, natural lakes. Terrestrial areas adjacent to the river-bed become inundated and their ecosystems destroyed. Soils and various forms of organic matter pass into the flooding water, while additional dissolved and suspended matter accrues from in-flowing rivers so that initially the waters are even more turbid than a river in flood. At least temporary, the impounded water takes its character from the materials which dissolve in it on contact, but during this early period, the distribution of chemicals ii unstable, varying with the pattern of the movement of the water.

The most dramatic feature of the ecological change is the initials ecological effect on the riverine fauna. Fishes appear most affected. Both vertebrate and invertebrate bottom dwellers become subjected to greater depths than those to which they are accustomed and the niches which form their habitats are destroyed in the turmoil. Those with filter feeding mechanisms find they cannot function satisfactorily without moving water. The highly turbid water obscures the sunlight needed for photo-synthetic processes in the medium. The terrestrial vegetation, flooded in an aquatic environment for which they are physiologically unprepared die and undergo decomposition. In the process, dissolved oxygen is used up and its concentration in the water may be seriously affected. Riverine organisms which cannot cope with, or escape from the new conditions, perish and quite often there are dead floating fish in newly inundated areas.

Clearly the major changes in the physical and chemical environment occur shortly after the creation of a new lake, but even so it takes many years for the system to take on a stable character of its own. This is particularly true for biota and there can be a long period when, ecologically, the system is neither typical of a lake nor of a river. Before a stable community is developed, there can be a succession of biological changes. These changes reduce the resistance and resilience of fauna and flora to climate change.

The accumulation of water to form a lake results in the permanent loss of some of the riverine resources but these can be offset by the acquisition of new resources, including biological ones. The expansion of water area and the new physico-chemical conditions in the water following impoundment inevitably lead to new biological features. Unfortunately, not all of them are desirable. Some of the plankton and algae, generally pelagic, find the expanded system suitably provided with nutrients and they increase rapidly.

It is quite common in such circumstances to have blooms, especially of the blue-green algae. This was true of the four large African man-made lakes, Kariba, Volta, Nasser and Kainji. Although soon after the inundation plankton was not abundant there was an algal bloom of *Microcystis* and *Anabaena* on Kainji. On Lake Nasser, *Volvox* was dominant on the new lake community, and on Volta Lake there were often masses of floating *Microcystis* and *Volvex* in the first years after the closure of the dam. To a large extent these algal blooms do not seem to have resulted in serious mishaps. Zooplankton generally was less abundant initially. On Volta, *Bosmina, Filinia, Karatella* and *Ceriodaphina* were among the most common. As the impoundments gradually settled down and the nutrient levels altered, the production rate improved and distributions became more balanced (Obeng, 1979).

Similarly, especially in tropical areas, there is usually an increased invertebrate population mostly in the shallower littoral areas. Thus on Volta, samples generally were rich in nymphs of dragonfly, caddisfly, and mayfly. *Povilla adusta* was particularly evident (Figure 4).

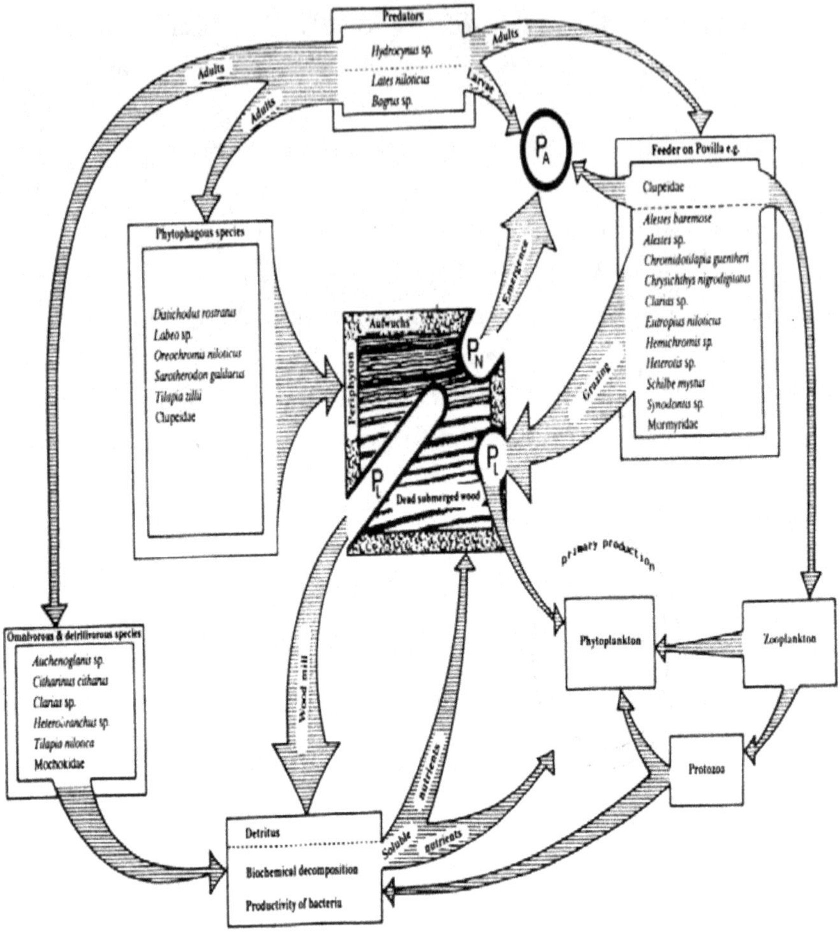

Figure 4: The role mayfly *Povilla adusta* (PA= adults, PN = nymphs, PL=larvae) in the possible chain of feeding strategies of African fish. *It is extremely difficult to classify the African freshwater fishes into rigid feeding habits as dietary changes occur at different ages and seasons of the year (Lauzanne 1988). The mayfly Povilla fits into the feeding strategy very probably according to the draft scheme (from Tobias & Lelek 1986)*

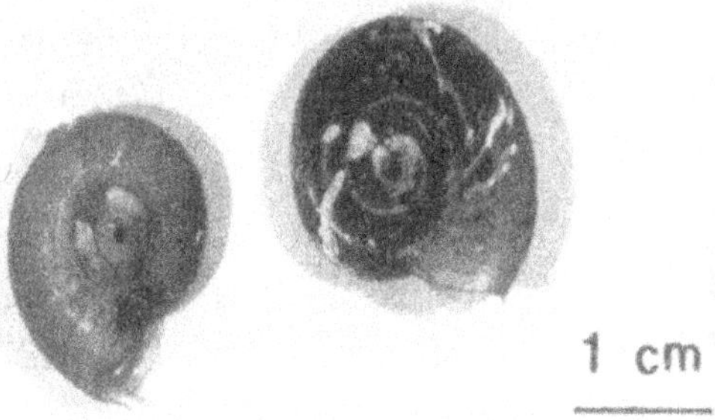

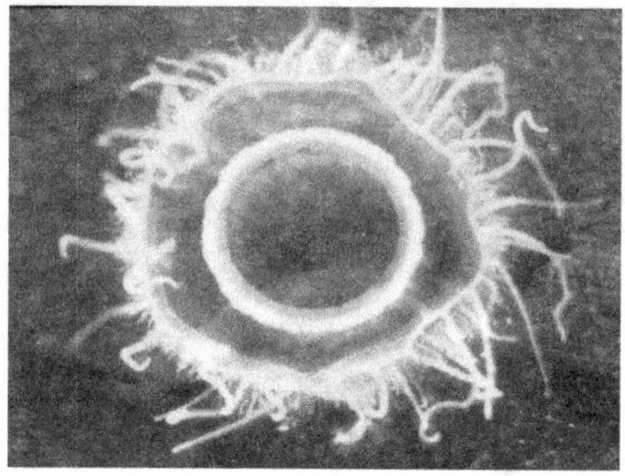

Photo: *Biomphalaria pfeifferi* is the intermediate host for the parasitic flatworm *Schistosoma mansoni,* the transmitter of intestinal schistosomiasia

Photo: After the initial filling of the man-made lakes, mass development of fresh water medusa *(Limnocrida tanganyicae*) frequently occurs. *In the words of the local fisherman "the lake is receiving eyes". The jelly-fish is about 25*

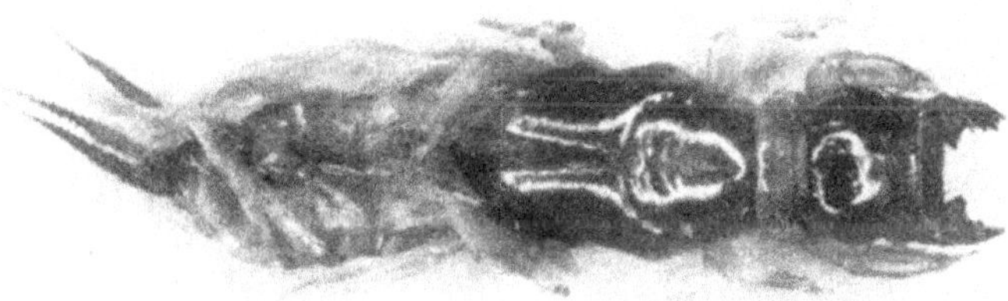

Photo: Adult nymph of *Povilla adusta. The strong jaws which the insect uses to dig its passages, even through hard wood, stand out clearly. Scale. 1 point – 1 mm*

There were also larval forms of mosquitoes including Mansonia Africana, as well as aquatic coleoptera and hemiptera. Kainji and Nasser also had larval mosquitoes. In the bottom samples in these areas, the fauna was represented by larval chironomids and oligochaetes. Under these conditions, the most favoured invertebrate fauna are the mollusks which, under tropical conditions and with an abundant supply of submerged and decaying vegetation, thrive and multiply to produce large populations. On the Volta dam, Bulinus species appeared and were well established within a year of the closure of the dam. Pila africana and other mollusks were also present. In addition to Bulinus globosus, which was also on Lake Kainji, Biomphalaria pfeifferi formed part of the snail community on Lake Kariba.

As the expanded water volume and the various forms of nutrients encourage an increased productivity in the various groups so the fish community also has a boom in production. With the increase in plants the herbivorous fishes do well. Small fishes such as clupeids thrive and therefore the larger piscivorous fishes feed well. The groups of surviving fish species reach huge sizes, reproduce at a high rate and the fish yield is enormous. On the Volta Lake, from the previous paltry catch on the river, within four years of closing of the dam the fish catch had multiplied several times to reach an annual conservative estimate of 60.000 metric tons. With the gradual stabilization of the lake conditions and reduction of nutrient levels, the catch later levelled to around 4.200 metric tons, which is still a good catch. Similar fish increases with reasonable stabilized catches have also been recorded on the other dams (Obeng, 1979).

The changes in the riverine system caused by dams and diversions can considerably reduce the resistance and resilience of aquatic life to climate change. The main ecological consequences are summarized as:

- Dams reduce river levels. By diverting water for power, dams remove water needed for healthy instream ecosystems. Stretches below dams are often completely dewatered.
- Dams block the rivers, prevent the flow of plants and nutrients, and impede the migration of fish and other wildlife. Fish passage structures can enable a percentage of fish to pass around a dam, but multiple dams along a river make safe travel unlikely.
- Dams slow rivers. Many fish species depend on steady flows to flush them downstream early in life and guide them upstream years later to spawn. Stagnant reservoir pools disorient migrating fish and significantly reduce the duration of their migration.
- Dams alter water temperatures. By slowing water flow, most dams increase water temperatures during the water storage phase in the reservoir. During the water release phase, dams decrease temperatures by releasing cooled water from the reservoir bottom. Fish and other species are sensitive to these temperature irregularities, which often destroy native

populations.

- Dams alter timing of flows. By withholding and then releasing water to generate power for peak demand periods, dams cause downstream stretches to alternate between low water or no water and powerful surges that erode soil and vegetation and slow or strand wildlife. These irregular releases destroy natural seasonal flow variations that trigger natural growth and reproductive cycles in many species.
- Dams fluctuate reservoir levels. Peaking power operations can cause dramatic changes in reservoir water levels, which degrade shorelines and disturb fisheries and bottom dwelling organisms;
- Dams increase oxygen levels in reservoir waters. When oxygen deprived water is released from behind the dam, it kills fish and vegetation downstream (Anon, 1999).
- Dams hold back silt, debris and nutrients. By slowing flows, dams allow silt to collect on reservoir bottoms and bury fish spawning habitats. Silt trapped above dams accumulates heavy metals and other pollutants. Dams also trap gravel, logs and other debris, eliminating their uses downstream as food and habitat.
- Dam turbines cut up fish. Following current downstream, fishes are drawn into and cut by the power turbines.
- Dams increase predator risk. Warm, murky, reservoirs often favour predators of naturally occurring species. In addition, passage through fish ladder or turbines injure or stun fish, making them easy prey for flying predators, such as gulls (Anon, 1999)
- Finally, dams dewater the estuarine zone along the coast of Cameroon. By dewatering this zone salinity conditions are altered in the mangrove swamp ecosystem. Fresh water inflows are critical in maintaining the proper concentration of salt in brackish water of lagoons and coastal wetlands where mollusk and other organisms dwell. Salinity may be one of the most important factors in determining shell fish productivity. The most sensitive are prawns, oysters and mussels (Chiras, 1991).

Current Problems of Man-Made Lakes

An obvious ecological change is the appearance of weeds: Sometimes the plants are on rivers, but artificial impoundments appear to be favourite habitats and weeds invade them very soon after their formation. In the early periods of the African dams, there was a dramatic appearance and extensive growth of *Salvinia nymphellula, Azolla africana* and other floating plants. There were also submerged plants, mostly *Ceratophyllum demersum.* Along the shores the conditions were conducive to the growth of emergent and marginal plants which included grasses like *Vossia, Polygonum* and *Alternanthera.* Water hyacinth, which has been a nuisance plant on a number of rivers in parts of the world, rapidly colonized and flourished on new impoundments in Asia and South America. Among other things these weeds interfere with water flow, obstruct passage boats, make fish harvesting by trawling difficult, and also provide shelter and satisfactory habits for the breeding of invertebrates including the disease host snails and mosquitoes. Larval forms of *Mansonia* species are reported to pierce *Pistia* plants for air in the breeding areas. Along the weeded shores, conditions are also suitable for tsetse fly to breed and there was an increase in trypanosomiasis around Lake Kariba. *Glossina* species have been caught in the vegetation along the Volta Lake (Obeng, 1979).

Health problems can arise as a result of biological changes around new lakes: A good example is provided by schistosomiasis. In tropical endemic regions, schistosomiasis invariably becomes established in new man-made lakes as the changed ecology support the host snails. Infected people foul the water with the schistosomiasis eggs because of inadequate domestic water supply and waste disposal facilities and transmission sites develop in the areas where the people have contact with the lake. The cycle is completed and in no time infection spreads. This has been the story of Volta and Kariba and to a Lesser extent of Kainji. On Lake Nasser the problem is especially acute in the irrigated areas. Both urinary and intestinal schistosomiasis occurs but each is closely linked with the establishment of the specific host snail.

There are a variety of other effects associated with man-made lakes though not all are seen in Africa and some more particularly associated with smaller impoundments. For example, there are the social aspects of man-made lakes resettlement and the effect of flooding on the economy of cultures. Other effects of impounding water include local climatic changes in temperature, wind strength and direction which, in turn, can affect wave action on the shore erosion. In certain areas, the stresses from the large accumulation of water may increase the tendency to seismic activity as suggested for the Kariba and Volta. There could also be an impact on ground water levels but this is yet to be fully studied. Both upstream and downstream of dams the habitat and breeding places of wildlife and waterfowl are often destroyed. The reduction of water flow itself causes an ecological change downstream – the invertebrate fauna either increases or diminishes according to the change. The clam *Egaria* was once abundant in an area of Volta River but with the dam and the change in flow, the population has dwindled almost to disappearing. Fishes downstream are also affected and those which migrate upstream to bread can no longer do so when dams are put in their way. The loss in water quantity downstream can interfere with agriculture which is dependent on the river (Goldsmith and Hildyard, 1984).

Use of Dams to Alleviate Negative Effects of Climate Change

Connectivity is an important feature of riverine ecosystems, as it can help preserve flow regimes, promote ecological integrity, and allow migrating aquatic fauna to move between different habitats at various life history stages. Connectivity becomes important as the effects of climate change increase. It will enable aquatic fauna to migrate to more suitable habitats as climate belts shift. Unfortunately, this conflicts with other development goals such as the construction of dams for agricultural purposes and the development of hydroelectricity.

Some development effort in the face of climate change seek to rescue drying wetlands by adding to water tables below, while on the contrary others seek to drain excess water from flood plain habitats for different reasons – interbasin transfer. Interbasin transfer of water has been proposed as a measure to arrest the shrinkage of Lake Chad. In the same area interbasin transfer of flood waters has also been proposed as a measure of rationalizing available water in time and space. Some experts have proposed interbasin transfer as a possible adaptation to climate change in Africa and South America (Pringle, 2000; Smith and Lenhart, 1996). However, interbasin transfers result in the intermixing of diverse faunal communities that were previously isolated, and may have unforeseen effects on native communities (Davis *et al.,* 2000). In addition, tropical rivers for which many of these projects have been proposed are particularly vulnerable to the negative effects of interbasin transfers (Pringle, 2000).

Flow conditions below dams could possibly be manipulated to relieve some of the negative effects of climate change, for example, by releasing bursts of water to simulate flooding events in flood plains and their associated wetlands (Middleton, 1999; Vaselaar, 1997) or releasing water from multiple reservoir layers to control temperature (McCarty and Zedler, 2002). However, although these actions may help alleviate some of the negative effects of climate change on systems already impacted by dams, erecting dams solely to control water flows would cause far more damage to the ecosystem than the original alteration in flow regime caused by climate change. Retaining water and altering its natural flow can lead to large changes in aquatic and terrestrial habitats, both above and below dams (Goldsmith and Hildyard, 1984; Baron *et al.*, 2003).

Other modifications of riverine systems involve the construction of obstructions such as levees and dykes to prevent water from flowing laterally over river banks during high flows – in this way, land adjacent to rivers can be developed with less risk of flooding. However, these barriers disrupt connectivity between rivers and valuable floodplain habitats, which serve as refugia and spawning grounds for aquatic fauna, and often sites of high biodiversity (WWF, 2003).

Planning and Management Guidelines

Despite the attendant problems dams are useful and necessary for development. For example, hydroelectric power production does not pose the type of pollution problems leading to global warming. Dams have added use of water storage, fish production, transportation and tourism. To ensure that the adverse impacts do not destroy their usefulness, there is in each case the need for long-term monitoring of physical, chemical and biological changes that can shape management decisions.

For most large dam projects, the government sets up an Authority to be responsible for aspects of the development. Such an Authority needs to be able to organize inter-disciplinary cooperation and to be effective it should have a specific multidisciplinary supporting unit with defined competence for the holistic appreciation of the complex problems of dams and the ecological events which produce the adverse environmental effects that they have.

Ideally, there should be a scientific unit to be responsible for monitoring the lake in order to advise the managing authority on the ecological situations that arise. There is now a great deal known about the physical, chemical and biological effects of dams which is why there is need for multidisciplinary approach to their planning and construction. Could engineers, for instance, advise on how to have dams with weed-free shores? What precautions should be taken to prevent the explosive growth of submerged quatic plants? Can agriculturealists ensure better use of fertilizers in catchment areas in order not to over-fertilize lakes and ponds? If lakes must be fished, what would planners advise in order to minimize human contact?

Health authorities and social workers need to ensure safe waste disposal systems. Ecologists and biologists must monitor contaminating factors. Organized long-term research, surveillance and assessment of ecological changes are essential. Parents and children have to cooperate in preventing contamination of shores. Social and community workers must foster community involvement.

Making a dam safe is a complex procedure. It involves not only considerable expense but the will to take precautions to make it environmentally safe. Only when this fundamental responsibility has been accepted and measures implemented to ensure a correct environmental appraisal can a decision on the building of a dam be made.

As dams age attention should be paid to restoration and repair options. Apart from the need to install aquatic animal passage facilities there is also the need to prepare sediment management plans. The steps needed are:

- Determine the reservoir sediment characteristics, including volume, particle size, distribution, density and environmental quality.
- Investigate the predevelopment and current geomorphology of the river basin, including the river channel upstream and downstream.
- Inventory the existing infrastructure around the reservoir, along the downstream river channel and along the upstream river channel that can be influenced by discharge from the reservoir and storage volume of the reservoir.
- Determine the feasible range of the sediment management options and formulate specific alternatives.

Conclusion

The building of large dams is a subject of great controversy. The basic point at issue is whether their benefits outweigh their costs. The benefits are considerable but some of the problems raised in this chapter were not foreseen when the projects were being planned. Now they can not be ignored. Often it will be found that a series of smaller projects would provide most of the benefits at lower investment cost and greatly reduced social and environmental impacts. The construction of large dams reduces the resistance and resilience of aquatic life to climate change.

References

Abell, R.; Thieme, M; Dinerstein, E.; and D. Olson (2002) A source book for conducting biological assessments and developing biodiversity visions for ecoregion conservation. Volume II: *Freshwater Ecoregions,* WWF, Washington D.C. 201p.

Anon.(1999) American rivers. *Journal of the International Erosion Control Association, Vol. 6,* p. 40-42.

Baron, J.; Poff, N.; Angermeier, P.; Dahm, C. *et al.* (2003) Sustaining healthy freshwater ecosystems. *Issues in Ecology Vol. 10*, p.1-6.

Chiras D. (1991) Environmental science. The Benjamin/Cummings Publishing INC., California, 549p.

Davies, B. Snaddon, C. Wishart, M.; Thom, M.; Meador, M. (2000) A biogeographical approach to interbasin water transfers: applications for river conservation. In: Boon, P.; Davis, B.; Petts, G. (eds) *Global perspectives on river conservation: science policy and practicals.* John Wiley and sons, INC. West Sussex.

Goldsmith, E. and Hilyard, N. (1984) The social and environmental effects of large dams. Vol. 1 Waderbridge Ecological Centre. Cambridge, 287 p.

McCarthy, J. and Zedler, J. (2002) Restoration ecosystem. In: H.A. Mooney and J.G. Canadell (eds). *The earth system: biological and ecological dimensions of global environmental change, vol.2.* John Wiley and Sons INC., Chichester.

Middleton, B. (1999) Wetland restoration, flood pulsing, and disturbance dynamics. Wiley, New York, 388p.

Molinier, M.; Sighomnou, D. and Sigha, K. (200) Perturbation du milieu naturel du Yaere dans le Nord Cameroun: changements climatiques ou action anthropiques? *Cameroon Geographical Review Vol. 14,* p. 155 – 170.

Ngwa, J.A. (1979) A new geography of Cameroon. Longman Group Limited, London, 151 p.

Obeng, L. (1979) Man-made lakes and their problems. Lectures presented at *6th FAO/SIDA Workshop on Aquatic Pollution in Relation to the Protection of Living Resources – Nairobi/Mombassa.* F.A.O. Rome, p. 138 -142.

Olivry, J.C. (1978) Suspended solid transport load in Cameroon rivers. *Science and Technology Review, Vol. 1,* DGRST, Yaounde.

Pringle, C.M. (2000) River conservation in tropical versus temperate latitudes. In: P.Boon; B. Davies; G. Petts (eds). *Global perspectives on river conservation, science, policy and practice.* John Wiley and Sons INC., West Sussex.

Smith , J. and Lenhart, S. (1996) climate change adaptation policy options. Climate Change Research, Vol. 6. p. 193 -201.

Vaselaar, T.R. (1997) Opening the flood gates: the 1996 Glen Canyon Dam experiment. *Restoration and Management Notes, Vol. 15,* p.119 – 125.

WWF (2003) Buying time: a user's manual for building resistance and resilience to climate change in natural systems. World Wide Fund, Washington D.C.

Chapter Ten

Climate Change and Management of Protected Areas

Summary

There is a growing recognition that natural ecosystems, both large and small, could provide a suite of ecosystem services related to climate change, ranging from protection against immediate physical impacts (rising temperatures, unstable climates, rising sea level) to providing additional insurance against the predicted instability of agriculture, fisheries and water resources. The chapter describes the general ecological and biodiversity features, distribution per biogeographic region of both the biodiversity and protected areas, and appraises the management. It further examines the significance of particular impacts and how protected areas might relief symptoms. The main issues examined include: disaster mitigation covering floods, droughts, landslides, and coastal erosion; fires, biodiversity; water security; the need for rapid crop adaptation; and food security, including the potential of crop and fish stock failure. Finally, some recommendations are made on how protected area management can contribute to the wider efforts at mitigating climate change impacts, and incidentally increase support and resources for management.

Key Words: ecological benefits, economic benefits, protected areas, climate change impacts, managing to relief impacts.

Introduction

Available literature points to the fact that well-designed and managed protected areas may ameliorate some of the problems caused by climate change, with benefits well beyond the boundaries of the protected areas. These potential benefits are predicted from a number of assumptions about the way in which natural ecosystems will respond to changing climate. This chapter seeks to describe the general ecological features and biodiversity potentials, the distribution of protected areas per biogeographic region, appraises the management and outlines some of the known and expected benefits that protected areas offer to resilience against climate change. It identifies the scope for maximizing these benefits.

Ecological and Biodiversity Potentials

The diversity of relief and latitudinal extent of Cameroon has produced several micro-ecological climates that are reflected by a rich and varied biodiversity. One can encounter several ecological niches over very short distances in the mountain and sub-montane areas. Figure 1 presents the various vegetation zones, and the zone of ancient and recent destruction and disappearance of the rainforest. The overlay of climate change impacts and human impacts has contributed to the savannization of this zone. This also is evidence of the southward shift in vegetation types, that is, the recession of forest types to the advantage of invasive weedy species, graminae, fire-tolerant and more xerophytic species. Savanna of post forest is spreading their antennae into the once humid forested areas. The transition is from humid forest to tree and shrub savannah, shrub savannah, shrub steppe and grassland savannah. These derivatives, however, reflect the intensity of the human impact on the environment, edaphic factors, altitude, latitude and aspect. Future climate change will exacerbate both the anthropogenic or human-related impacts and the climate-related impacts. This will endanger several ecosystem goods and services. These are important potentials to be protected and sustainably managed. Table 1 and 2 present some wildlife data for the country.

The country possesses 297 species of mammals, 848 species of birds. 300 species of amphibians, 9000 species of plants (Gartland, 1992), 29 species of primates and 39 swallowtail butterflies (USAID. 1991'). Some 156 species of plants are endemic (Gartland, 1992). Much of Cameroon's potential wealth thus lies in its natural resource base. This resource base includes forest, biodiversity, and soil and water resources. This biodiversity provides great potential for eco- tourism, for timber and non-timber forest products. The forests are also of great scientific value for the world community.

If we consider only the vertebrates: Cameroon has 2 1% of the fish species found on the African continent, 48% of mammal species. 50% of batrachian species, 54% of bird species and 30 to 75% of reptiles. The presence of these animals is closely linked to the vegetation in which they live. There are many ecosystems in Cameroon. We can distinguish three main Zoographic zones: the forest, Savannah and Sahel zones.

Mammals are because of their size the most spectacular element of the wildlife population and contribute enormously to the tourist resource base of the country. There are five families of primates. These are all forest species except for three species of monkeys and one type of bush baby. 34 species of carnivores are present. Although the lion and the jackal are still well represented in the national parks, the panther, cheetah, lynx, bush cat, golden cat, hyena and light-coloured fox have all become very rare. Elephants, which used to be very numerous, survive in the sparsely populated areas, but it is common to meet the Savannah elephant in the national parks. The rhinoceros has seen its numbers diminish drastically in recent years. The hippopotamus, originally found in all Cameroon rivers, has disappeared from many of them. Buffalos and antelopes are still well represented. The country also possesses the three suidae of Ethiopia; Warthog, river hog and the giant forest hog. The order of rodents comprises many species. Bats are numerous.

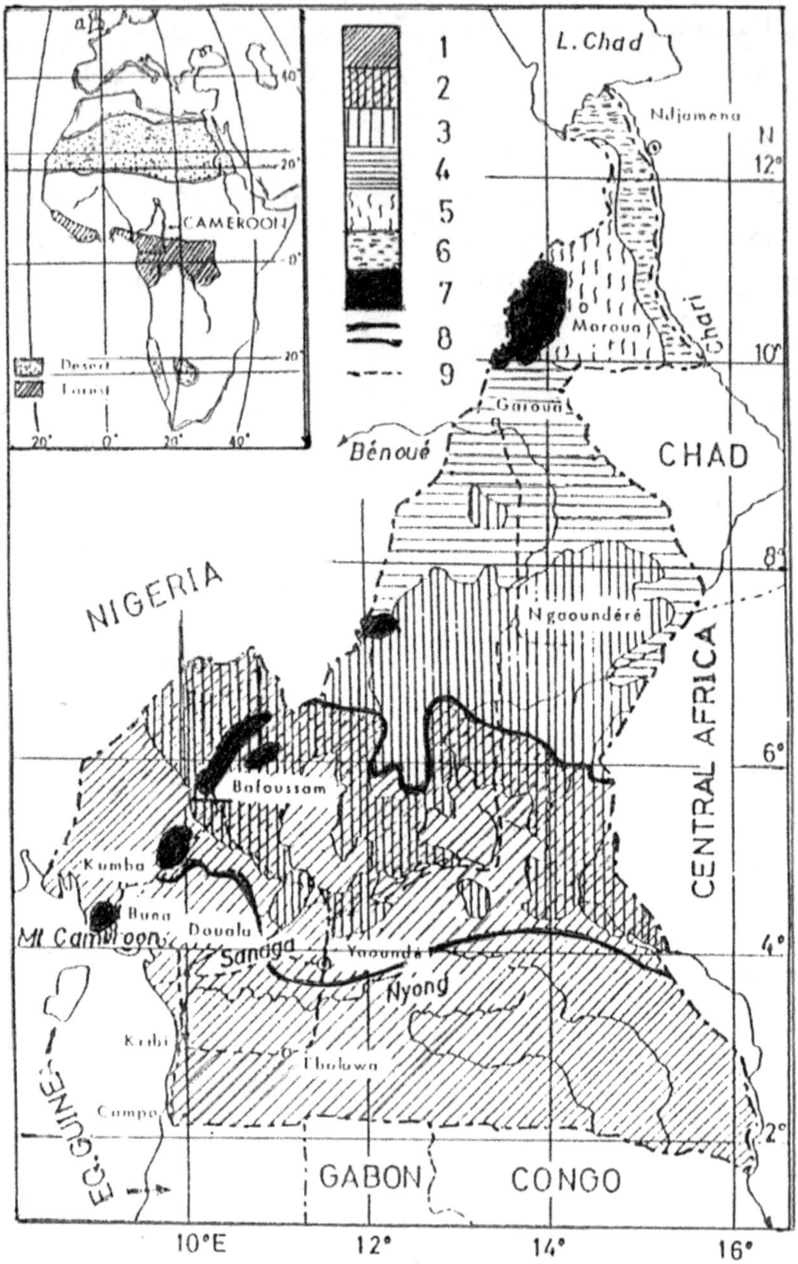

Figure 1: *Vegetation zones (after Letouzey, 1979; Loung, 1973) of Cameroon and main survey routes. 1: Dense moist forest, 2: Savanna of post forest, 3: Shrub savannah and grass savannah, 4: Tree savannah and savannah wood land, 5: Tree and shrub steppe, 6: Seasonally flooded tree and shrub steppe, 7: Mountain and subalpine formations, 8: Zone of ancient and recent destruction of dense moist forest, 9: Main survey routes.*

Photo: *Thorn bush savannah: Chad Plain*

Photo: *Dry Sudan savannah biogeographic zone*

Photo: *Sahel savannah biogeographic zone: Diamare*

Photo: *Wooded savannah biogeographic zone: Benue*

Photo: *High savannah in the Bamileke plateau*

Photo: *Montane forest on Mount Cameroon*

Photo: *Dense rainforest*

Photo: *Exploitation of the rainforest*

Stemming from the diversity of ecosystems, the bird population is extraordinarily rich with, to date, 942 species having been identified. Numerous sea birds are seen at the coast in the dry season (grills, petrels, skuas, puffins). The coastal lagoons and mangrove swamps harbour cormorants, sterns, herons, egrets, sandpipers, godwits, curlews, and plovers. The forest region houses the largest and richest bird population. Tauracos, parrots, guinea fowl, black partridge, pigeon, shrike and magpie are very rare. Taurocos have been classified as endangered due to habitat loss. In the Savannah eco-climate are large birds (jabine, black Abyssinian horntail, marabou, crowned crane, ibis, stork, heron, bateleur eagle, hoopoe, vulture, kite). The greatest gatherings of birds here can he seen around water sourceses during the dry season. The Sahel has considerable populations of ostriches, bustard, guinea fowl, black partridge, stone curlew, lapwig, nightjar turtledoves and sparrows. Some are sedentary while others are migratory. The ubiquitous species include: the magnificent black and white bald buzzard, cattle egret, pygmy kingfisher, jacana, cormorant, pagoda cock and omnipresent magpie.

The country is rich in reptiles. The crocodile, turtle and tortoise population is well known. The snake, lizard and chameleon population have not been well studied and identified. Three of the crocodiles existing in

Africa are present: Nile crocodile, long-nosed crocodile and snib- nosed crocodile. Among the tortoises and turtles are 5 types of sea turtles, 10 types of freshwater turtles and 4 types of tortoise. More than 150 species of snakes are found in the whole country. Fifteen of them are considered dangerous to man. With about 300 species of amphibians, Cameroon is the richest country in amphibians. The amphibians are composed of those of the lowland forest, secondary formations, mountains and savannah ecological niches. The most remarkable life forms in the forest are the goliath frog and hairy frog. The goliath frog is 30cm long and weighs 2.4kg and is the largest in the world. Most of these are endangered by habitat loss.

Man's aggressiveness towards nature is a well-known phenomenon all over the world. Cameroon is no exception. The population explosion is the main cause of the drop in the wildlife population. The actions of man on the environment, and thus on the wildlife, are harmful. This is particularly remarkable in the rain forest, where agriculture and lumbering have greatly modified the environmental structure and ecosystem functions.

Table 1: *Comparison of number of species of main flora and fauna represented in Africa and Cameroon*

	Group or class	Cameroon	Africa	% Cameroon / Africa
Flora	Phanerogames	8.000	45 to	16 to 17%
	Cryptogams		50.000	
	Pteridophytes	264		44 to 52%
			500 to 600	
	Fish (1)	530	2510	21%
Fauna	**Batracians (2)**	200	400	50%
	Crocodiles	3	4	50%
	Tortoise	12	40	30%
	Reptiles (3) Lizards	160 – 170	300 – 400	40 to 56%
	Snakes	160 – 170	250 – 300	53 to 68%
	Birds(4)	942	1.738	54
	Mammals(5)	126	263	48%

Source: Encyclopaedia of the United Republic of Cameroon, Vol. 1(1979) Les Editions Africaines

Table 2: *Comparison of mammals represented in the world, Africa and Cameroon*

Order	Family	Sub-family	World	Africa	Cameroon
Insectivore:	Erinaceidae		14	6	1
	Potamogalide		3	3	1
Lagomorphe	Leporidae		59	7	1
Rodents	Sciurisae		285	25	11
	Anomaluridae		9	9	7
	Thryonomyidae		6	2	2
	Hystricidae		20	4	2
	Pedetidae		2	1	0
Pholidotes	Manidae		7	4	3
Primates	Lorisidae		5	2	2
	Galagidae		5	4	
	Cercopithecidae		49	28	16
	Colobidae		24	7	3
	Pongidae		3	2	2
Carnivore	Canidae		35	11	4
	Mustelidae		42	7	5
	Viverridae		70	29	16
	Hyaenidae		4	4	2
	Felidae		37	10	10
Tubulident	Orycteropidac		1	1	1
Sirenian	Trichechidae		3	2	1
Hyracoide	Procaviidae		9	3	2
Proboscidian	Elphanitidae		2	1	1
Perissodactyle	Equidae		12	4	0
	Rhinocerotidae		5	2	1
Artiodactyle	Hippopotamidae		2	2	1
	Suidae		8	4	3
	Tragulidae		2	1	1
	Giraffidae		2	2	1
	Bovidae	*Hippotraginae*	6	6	1
		Ruduncinae	10	10	4
		Alcelaphinae	9	9	2
		Antilopinae	20	13	1
		Caphalophinae	14	14	9
		Neotraginae	7	7	3
		Madoquinae	4	4	0
		Caprinae	28	2	0
		Bovinae	16	1	1

Source: *Encyclopaedia of the United Republic of Cameroon, Vol. 1(1979) Les Nouvelles Editions Africaines*

Man has always altered habitats, but today, the demographic boom of the country, introduction of the cash crop economy, the needs of subsistence farming, commercial and subsistence hunting and organization of the human or cultural space place inordinate strain on habitats essential to many species (see Tables 3, 4 and 5). Alteration of habitat is the most significant single factor in extinctions. Habitat is destroyed by human civilizations spreading into fields, forests, oceans, mountains, riparian areas and water ways, roads, strip mining, dams, housing projects, airports, farms and cities usurp wildlife habitat. Industrial activities especially along the coast have altered several ecological niches.

Table 3: *Crocodiles found in Cameroon and their habitats*

Scientific name	Maximum length	Habitat
Crocodylus niloticus	6 metres	Ubiquitous
Crocodylus cataphractus	4 metres	Costal zone
Ostrol / Emus tetraspis	2 metres	Rainforest

Source: *Cameroon Encyclopaedia, Vol. 1, 1977, ENA*

Table 4: *Principal land tortoises of Cameroon and their habitats*

Family	Scientific name	Maximum length	Habitat
Testudinidae	*Geochelone sulcata*	80cm	Sahel
	Kinixys belliane	20cm	Savanna
	Kinixys erosa	30cm	Forest
	Kinixys homeana	20cm	Forest
Pelomedusidae	*Pelomedusa subrufa*	25cm	Savannah
	Pelusios gabonensis	30cm	Forest
	Pelusios castaneus	35cm	Savannah
	Pelusios niger	30cm	Savannah / Forest
	Pelusios carinatus	25cm	Forest
	Pelusios adansonii	20cm	Sahel
Trionychidae	*Trionyx triunguis*	80cm	Savannah/Forest
	Cyclanorbis senegalensis	50cm	Savannah
	Cyclanobis elegans	50cm	Savannah
	Cyclanobis aubryi	50cm	Forest

Source: *Cameroon Encyclopaedia Vol. 1, 1979, ENA*

Table 5: *Main species of snakes and their habitat*

Family	Scientific name	Maximum length	Habitat
Elapidae	*Naja nigriocollis*	2.20 m	Ubiquitos
	Naja melanoleuca	3.10 m	Forest / Savannah
	Naja haje	2.50 m	Savannah / Sahel
	Naje katiensis	2.000 m	Savannah / Sahel
	Boulangerina annulata	3.00 m	Forest
	Pseudohaje goldi	2.60 m	Forest
	Dendroapis jamesoni	2.80 m	Forest
	Dendroapis polylepis	4.00 m	Forest
Viperidae	*Causu lichtensteini*	0.60 m	Forest
	Causus resimus	0.75 m	Forest / Savannah
	Caussus maculates	0.50 m	Savannah
	Bitis arietans	1.60 m	Forest / Savannah
	Bitis gabonensis	2.20 m	Sahel / Savannah
	Bitis nasicornis	1.20 m	Forest
	Echis carinatus	0.80 m	Savannah
	Cerastes cerastes	0.80 m	Sahel

Source: *Cameroon Encyclopaedia Vol. 1, 1979, ENA*

However, the main factor in the disappearance of the wildlife remains hunting in all its forms. The pressure exerted by hunting has considerably increased through the growing numbers and sophistication of firearms. Added to hunting, which often takes place in defiance of all regulations, are trapping activities and the use of pesticides. The current tendency is towards a drop in numbers of all synergistically interesting animal species and even to the rareness of certain species to such an extent that there is the danger of them disappearing altogether. Most of them are today only found in national parks, reserves and protected areas which themselves are poorly managed.

The elephant populations have declined drastically during the past decades. The killing of elephants by poachers to supply the ivory markets primarily of Japan, the United States and Europe is the main cause.

Elephants are also jeopardized by loss of habitat as agriculture reaches deeper into the wilderness. The elephant's complex social structure has been greatly damaged and in some areas it is not clear whether surviving elephants will be able to sustain their populations through reproduction. Other species may become affected because the elephant is a keystone species, one that shapes its environment. The changes that the elephant brings to its environment are beneficial to a variety of other Savanna species. Loss of the elephant could have a drastic impact on the tourist business, it may also affect the rapidity with which trees populate savannas emptied of elephants; potentially changing the types of species that can use the land.

Biogeographic Distribution Of Protected Areas

The distribution of protected areas per biogeographic zone and the surface area of each protected zone are presented in Table 6. There are five main wildlife reserves covering a land surface area of 575,000 hectares. These reserves are protected by government regulations and equipped with amenities for tourists. These include Waza, Benue, Bouba-Ndjida, Kalamaloue and Mozog-Gokoro reserves. These reserves are found in the Sudanian -Sahelian eco-floristic zone:

- Waza wildlife reserve was created in 1968. It has a land surface area of 170.000hectares. This is the best known and the most spectacular wildlife reserve in the country. The main animals found here include the lion, cheetah, elephant, hippotragus, waterbuck, hartebeest, cob, giraffe, wartlog and the panther. There are several birds such as the ostrich, heron, goose, pelican, egret and the vulture among others. Lodging camps for tourists are perched on the Southern slopes of the hills overlooking Waza town.
- Benue wildlife reserve has a land area of 180.000 hectares along the Ngaoundere-Garoua highway. Traversed by the Benue River, there are several hippopotami and crocodiles in the shallow riverbed. Other important animals include buffalos, derby elands, antelopes, lions and elephants. The reserve has two lodging camps.

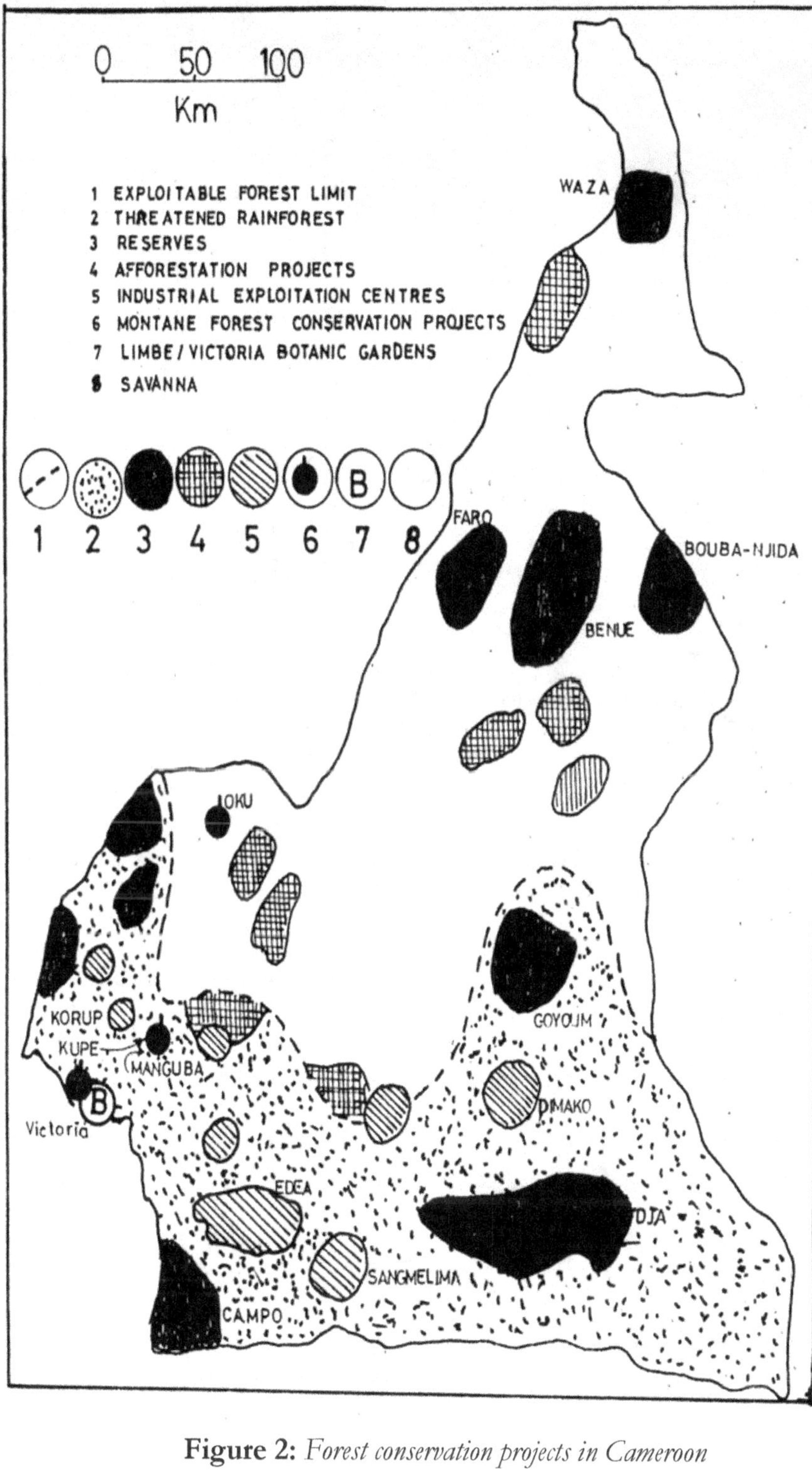

Figure 2: *Forest conservation projects in Cameroon*

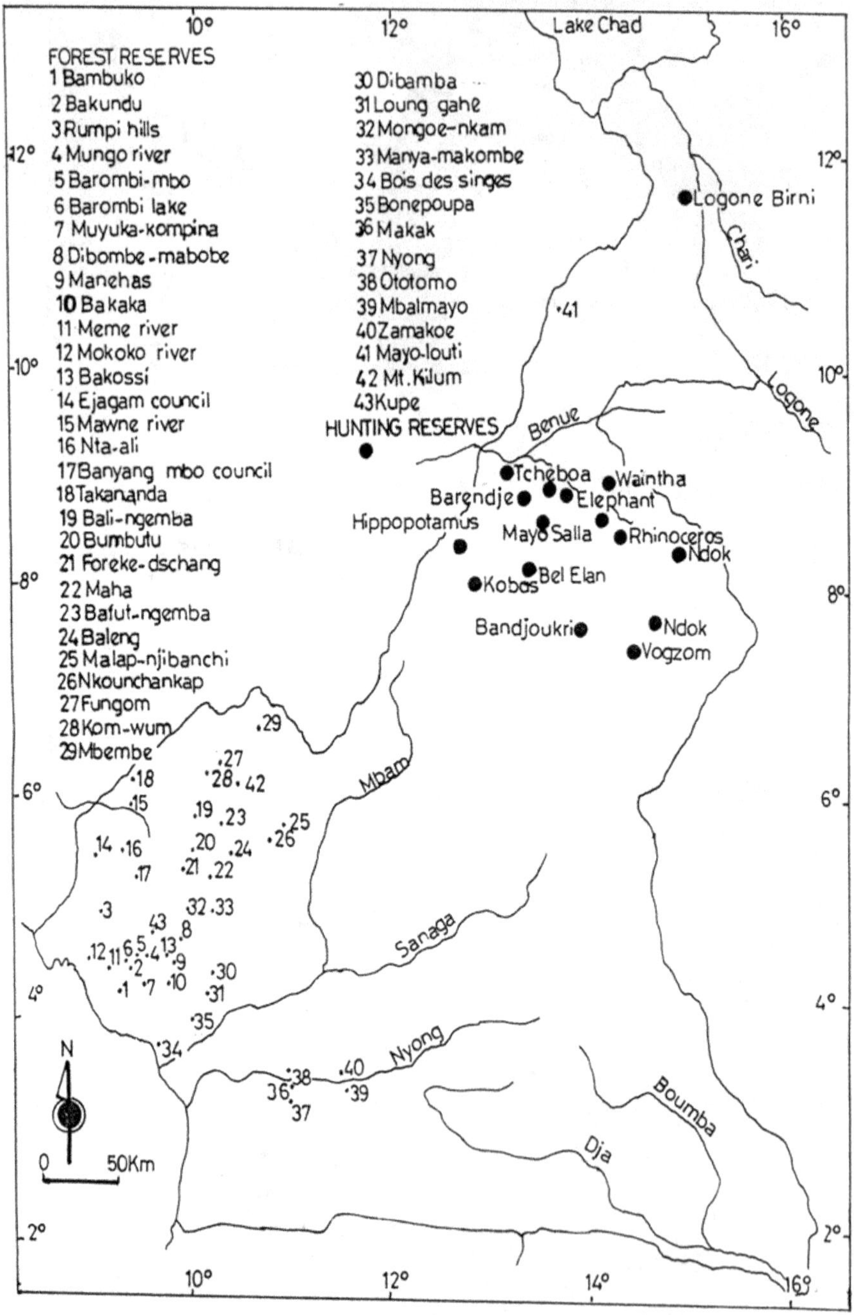

Figure 3: *Hunting and forest reserves of Cameroon situation as in 1992*

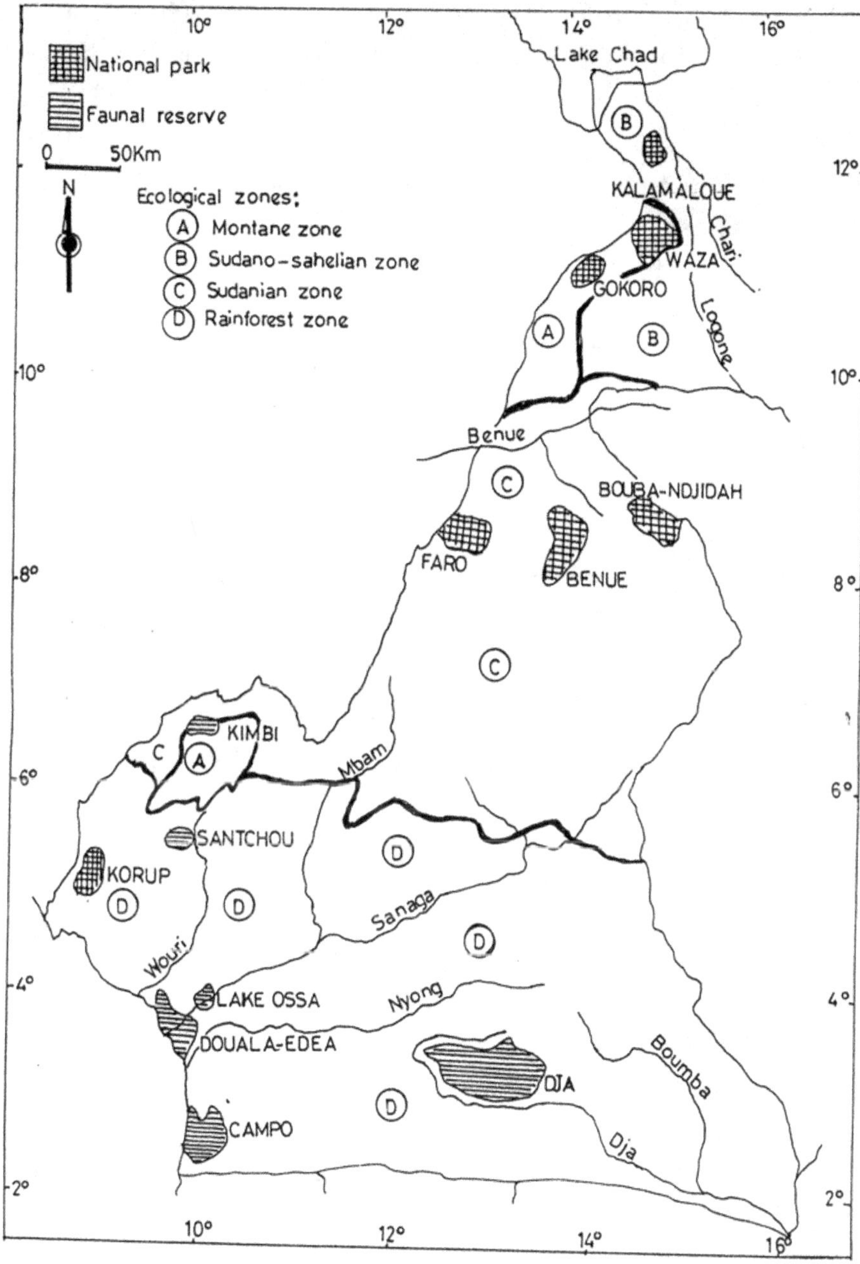

Figure 4: *National Parks and faunal reserves of Cameroon*

- Bouda-Ndjida wildlife reserve is located in the Cameroon-Chad frontier between latitudes 8° and 9°N. It occupies an area of 220.000 hectares and is the habitat for rhinoceros, giant eland, antelopes, lions and buffaloes.

- Kalamaloue reserve covers an area of 4,500 hectares near the border town of Kousseri. Like the Bouba-Ndjida and the Mozogo-Gokora reserve there are no lodging camps.
- Mozogo-Gokoro wildlife reserve is found near Koza pass. It has an area of 1400 hectares and is poorly developed in tourist facilities.

Apart from these reserves in the Savannah ecological climate, there are eight forest reserves that are rich in biodiversity. While the wildlife reserves are well developed in tourist amenities and routes the forest reserves do not have such facilities and therefore are not developed to receive tourists (Figure 2, 3 and 4)

Table 6: *Distribution of forest reserves per biogeographic region*

Forest Reserve	Area (hectares)	Biogeographic zone
Kalfou reserve	4000	Sahel eco-climate
Faro reserve	335,000	Sudan eco-climate
Pangar and Djerem	300,000	Sudan eco-climate
Kimbi reserve	5,000	Montane eco-climate
Dja reserve	520,000	Rainforest eco-climate
Douala – Edea reserve	160,000	Rainforest eco-climate
Campo reserve	300,000	Rainforest eco-climate
Lake Ossa reserve	4,000	Rainforest eco-climate

Within the rainforest region Korup is the only forest national park out of the seven national parks. The other six are located in the sudano-sahelian region (Faro, Benue, Bouba-Njida, Mozogo-Gokoro, Waza and Kalamaloue). Korup National Park contains Africa's oldest rainforest and is over 60 million years old with a high level of endemism. There are 1000 species of plants and 1300 animal species including 119 mammals and 15 primates. Of the total listed species, 60 occur nowhere else and 170 are listed as endangered. It was established in 1937 and would have largely disappeared by 2025 but for conservation work initiated by the World Wide Fund (WWF) for Nature. The conservation programme centres on a management area of 126,000 hectares plus a surrounding buffer zone of 300.000 hectares (Pearce, 1994).

In the Mount Cameroon region there are areas protected to maintain nationally significant landscape characteristics of the harmonious interaction of resident people and land while providing opportunities for public enjoyment through, recreation and tourism within the normal life-style and economic activity of these areas. (Table 7)

Photo: *Poaching is a major cause of the extinction of animal species. Villagers with a dead elephant after a fruitful hunt in a forest reserve in the South Cameroon rainforest. (after A. S. Neba, 1987)*

Table 7: *Protected areas of the Mount Cameroon Region*

No.	Protected Area	Area (Ha)	Year Established
1	Korup National Park	211.675	1937
2	Ejagham Forest Reserve	74.850	1934
3	Takamanda Forest Reserve	67.599	1934
4	Mative River Reserve	53.872	1951
5	Rumpi Hills Forest Reserve	45.843	1941
6	Bayang-Mbo Forest Reserve	42.606	1936
7	Nta-Ali Forest Reserve	32.982	1935
8	Bambuko Forest Reserve	26.677	1950

9	Southern Bakundu Forest Reserve	19.425	1940
10	Mokoko River Forest Reserve	9.065	1952
11	Bakossi Forest Reserve	5.517	1956
12	Meme River Forest Reserve	5.180	1951
13	Mungo River Forest Reserve	4.662	?
14	Barombi-Mbo Forest Reserve	855	1950
15	Botanic Garden Limbe	375	1892
16	Buea Fuel Plantation	300	1953

The Limbe Botanic Garden is bilaterally funded by the government of Cameroon and the United Kingdom. It is focused on tourism, environmental education, scientific research and conservation. There is also a herbarium with accommodation for visiting scientists. The Limbe zoological Garden is an annex of the Botanic Garden. Over the years the animal population in the Zoo has been declining due to the lack of proper management. There is an on-going project in the Zoo that seeks to create a primate sanctuary centre for indigenous endangered species. These two gardens are part of a wider project known as Mount Cameroon Project. The project seeks to assess areas of high biodiversity in the Mount Cameroon region and to prepare them for gazettement as forest reserves. The identified areas include: Etinde Mountain forest, Onge forest and Moliwe-mabeta forest. Mount Kupe Forest Reserve and Takamanda Gorilla Sanctuary are essentially game reserves with the objective of identifying and conserving endangered species and endemic species. Mount Kupe reserve covers 2000 hectares of land.

The montane forests of Cameroon to the west of the country are unique. They have one of the highest levels of endemism in the whole of Africa, particularly among birds and vascular plants. Mount Oku was identified the only extensive area of montane forest left in the Bamenda highlands. This protection reserve was established in the 1930s. By 1984 it needed critical conservation action. This was rendered by the International Council for Bird preservation (ICBP). The remaining forest covers 7000 hectares. Mount Oku

Forest Reserve is characterized by a large number of endemic species. Altogether 53 species of Montane forest birds are found in the Western highlands, 20 of these are true endemics. The Preuss's monkey *(Cercopithecus preussi)* is one of the most important mammals to occur. It is classified by the International Union for conservation of Nature and Natural Resources (IUCN) as endangered. It has only been recorded in a few patches of forest in Cameroon and Eastern Nigeria. Several small mammals have subspecies endemic to the Oku Forest. A few species of frogs *(Xenopus sp.)* are endemic to the forest. This protected area has good scientific and tourist potentials but lacks tourist amenities.

Protected Area Management

There is general shortage of personnel in the national parks and wildlife reserves. For Instance Dja Wildlife Reserve, which measures some 526,000 hectares, has only one conservator and three guards. In addition, nearly all protected areas do not have adequate operational means of transportation, let alone good communication. The guards are equipped with obsolete weapons and are short of ammunition whereas poachers have more efficient weapons (Besong and Ngwasiri, 1995). These problems certainly greatly constrain biodiversity and eco-tourism industry.

While it is apparent that biodiversity protection has gathered 'considerable momentum in the country in recent years, there is still a wide gulf between rhetoric and policy objectives, on the one hand and the reality of policy and project implementation on the other. Numerous environmental laws and regulations enacted in 1994 remain unenforced, programmes or projects poorly implemented, while measures coined to protect or rehabilitate ecosystems often impinge negatively on livelihoods at the village or local level or involve benefits which accrue mainly to local elites. The experience reveals the following problems:

- Insufficient attention to socioeconomic and cultural situation of local people whose livelihoods depend on resources found in parks and preserves;
- Attempts by development agencies or the state to protect or rehabilitate ecosystems in one particular area are often contradicted by other development measures which degrade the environment.

- Many laws and regulations governing land use in parks and preserves remain unenforced
- It has been usual of protected area status to be imposed on a specific area without prior consultation with the local population. This results in social conflict in and around these areas which are generally associated with four factors: the ways parks and preserves have been established; the process of land acquisition; the invasion and occupation of land in protected areas; and resource use in such areas by the local population.
- The piecemeal and partial character or non-implementation of many conservation policies and programmes, as well as the difficulties of sustaining positive initiatives through time, and replicating successful interventions. This is largely as a result of dependence on external aid and expertise.
- Lack of adequate consultation and clarification concerning the demarcation of preserves and parks as well as the limited capacity of conservation agencies to enforce the protected area status.
- It is not uncommon to find that certain local interests such as logging companies or lumber merchants have the necessary financial and administrative resources, as well as the connections with local administrative authorities, to obtain the documentation required by the forestry Law for timber exploitation.

It is now being realized by development agencies that the "conservationist" approach to biodiversity conservation has failed to come to grips with crucial social issues. In most protected sites technocratic formulas have often been imposed which generally ignore the socioeconomic and cultural situation of thousands of families whose livelihoods depend on the forest. This approach has provoked social conflicts which often undermine the possibility of implementing and achieving basic conservation objectives. Under such circumstances, non-enforcement of regulations becomes an explicit strategy of the state to reduce conflicts. Coupled with limited human and financial resources necessary for the administration of parks and protected areas, most reserve status exist only on paper.

Most important, the sustainable management of wildlife requires understanding that future threats will be driven simultaneously by global phenomena such as the greenhouse induced climate change, as well as local and regional resource management schemes. Climate change will fundamentally change ecosystems composition and functioning.

Managing Protected Areas to Relief Climate Change Impacts

Carefully designed and managed protected areas can help relief problems beyond the park boundary. In some cases protected areas are simultaneously a potential buffer against a particular climate-related problem and at risk from the same problem, creating a tension and the need for some tough decisions from protected area managers. This requires a lot of research and monitoring in order to learn about many of these impacts. Opportunities to use protected areas as buffers against climate change require two additional management approaches (WWF, 2003):

- *Planning to maximize the benefits of protected areas in relieving the symptoms of climate change:* The importance of assessing protected area management effectiveness must be recognized and climate change elements must be added to these assessments. This apart from enhancing the effective management of the protected area will also aid in planning future protected areas. Such assessments can be carried out at ecosystem or landscape level. Management actions can then be put into a national or regional context.
- *Using protected areas to help mitigate the effects of climate change:* Available literature shows that protected area managers are already using land and water resources to buffer against climate change without recognizing the connection. However, more systematic research is required to quantify beneficial impacts and to refine management interventions to maximize these benefits. Efficient management of protected areas can achieve the following:
    - *Flood:* The presence of natural vegetation generally reduces storm-related erosion and landslides. Deforestation in several areas has led to flooding (Calder, 2000; Kaimowitz, 2002).
    - *Drought:* Maintaining natural vegetation can provide an important insurance policy in areas prone to drought. Drought effects have been exacerbated by prior forest loss (WWF, 2003)

- *Coastal impacts*: Protected areas can play a role in disaster mitigation in marine and coastal areas. Sea level rise and increased storm damage puts coastal communities and small islands at particular risk. The disappearance of Cape Cameroon before our eyes is clear evidence. Building physical barriers against the rising sea is technically difficult and colossally expensive for a poor country. This stimulates a new approach to integrated management based on the mangrove ecosystem. A review of marine ecosystem services suggests that the mangrove forests are the most cost effective option. Unfortunately, the consequences of their destruction are already evident in Limbe, Douala and Cape Cameroon. The restoration of the mangroves is extremely important (Moberg and Ronnback, 2003; Field, 1999).

Unfortunately, this ecosystem is paradoxically seriously threatened by climate change. Protected area managers will be faced with a series of hard decisions, balance the benefits from mangrove ecosystem, the chances of its surviving climate change and the cost and benefits of protection trade-offs may be necessary.

- *Fire:* Climate change will add an additional element to the already complex relationship between fires and natural systems. Hotter, drier conditions tend to increase fire frequency resulting in changes in vegetation as more fire-tolerant species become common. Forest fires can directly affect protected areas as is often the case in the Kilum and Ejim Forest Reserves. (Macleod, 1986; Goldammer and Seibert, 1992). Training in community fire management is therefore an important part of management (Karki, 2002).

- *Biodiversity:* Protected areas aim at protecting biodiversity. But managers must simultaneously also design management strategies that will increase the resistance and resilience of the biodiversity to climate change.

- *Water security:* Protected areas can be managed to guarantee the quantity of water supply. Forests increase or stabilize water supply. Some forests may also increase net water flow from watersheds. Good quality water, free from sediments is also guaranteed. This provides an argument for the protection of montane cloud forest and upland watersheds in river basin systems.

- *Genetic stress:* changing climate will increase stress on both new and traditional crop varieties. There is need for crop breeding programmes to respond to this change. Climate change will cause a decrease in crop varieties (McCarthy *et al.,* 2001). Wild relatives of modern crops therefore have an important role to play in kick-starting the breeding of new strains under time pressure. Protected areas can be designed to protect sources of crop genetic material.
- *Food security:* Climate change will also disrupt agricultural systems and fisheries. Protected landscapes and seascapes provide a potential buffer in three ways (McCarthy *et al.,* 2001): by providing breeding grounds for commercial species, by providing wild foods to the poorest members of society in time of crop failure and preserving genetic resources needed for adaptive breeding.

Conclusion

From the above review, the management of protected areas should therefore focus on:

- *Maintaining and increasing large reserved areas:* Large unfragmented reserves are likely to be most useful in both resisting climate change and providing resilience within a landscape, that is, increasing core reserve areas, linking reserves and developing effective buffers are crucial first steps.
- *Plan protected areas with disaster mitigation in mind:* Protected areas can protect human communities from climate-related floods, landslides and droughts.
- *Recognize the role of protected areas in maintaining terrestrial food and water supplies:* Small changes in management can allow the natural vegetation in protected areas to help supply emergency food and forage, increase or stabilize water yields which are all likely to be in shorter supply in the future.

- *Site marine protected areas to maintain fisheries:* As fish stocks suffer from multiple stresses of over-exploitation, pollution and climate change, ways of maintaining breeding stocks become more important. Portions of the Cameroon's Atlantic Ocean need to be protected.

References

Asong, A. (2001) Forest degradation and forest reserve strategies in the South West Province. In: Dunlop, J, and Roy, W. (eds.) *Culture and Environment.* University of Strachdye/Univerisity of Buea.

Besong, J. and Ngwasiri, C. (1995) The 1994 forestry law and rational forest resource management in Cameroon. PVO – NGO/NRMS Cameroon Publication, Yaounde.

Biwas A. K (1992) Environment and sustainable development for Cameroon. Report of a multidisciplinary and multi-institutional mission on the environment, Yaounde.

Calder, I. R. (2000) Land use impacts on water resources. *Background Paper No.1. Electronic Workshop on Land-Water linkages in Rural Watersheds.* FAO, Rome.

CAMPER/USAID (1993) USAID Cameroon programme for environmental reform. CAMPER.

Field, C. D. (1999) Rehabilitation of mangrove ecosystem: an overview, *Marine Pollution Bulletin, Vol. 37*, p. 8-12.

Gartland, S. (1992) Cameroon. In. Sayer, J. (ed.) conservation atlas of tropical forests, Africa. Harcourt and M. Collins, London.

Goldammer, J.G. and Seibert, B. (1992) the impact of drought and forest fires on tropical lowland forest of East Kalimantan. In: J.G. Goldammer (ed.). *Five in the tropical biota-ecosystem processes and global challenges.* Springer-verlag, Berlin.

IUCN(1978) Threatened mammals of Africa. The IUCN Red Data Book. International Union of Nature and Natural Resources, Cambridge.

IUCN (1985) 1985 United Nations List of National Parks and Protected Areas. IUCN, Gland.

Kainmowitz, D. (2002) Useful myths and intractable truths: the politics of the link between forest and water in Central America. Centre for International Forestry Research, Bogar, Indonesia.

Karki, S. (2002) Community involvement in and management of forest fires in southeast Asia. Project Fire-Fight SE Asia. WWF, ICUN and European Union, Bogor, Indonesia.

Letouzey, (R. (1979) Vegetation. In: J.F. Loung (ed). *Atlas of the United Republic of Cameroon.* Jeune Afrique, Paris, p. 20-24.

Loung, J. F. (1973) Le Cameroun. Collection A. Journaux, Hatier, Paris. 96p.

Macleod, H. (1986) Conservation of Oku Mountain forest, Cameroon. *ICBP study Report No. 15.* Cambridge.

McCarthy, J.; Canziani, O.; Leary, N.; Dokken, D.; and K.S. White (eds.). Climate change 2001: Impacts, adaptation, and vulnerability. IPCC, Cambridge University Press.

MCP (1998) Report on the elaboration of pluralistic sustainable structure for the management and conservation of natural resources in the Mount Cameroon Region. MCP, Buea.

Moberg, F. and Ronnback, p. (2003) Ecosystem services of the tropical seascape: interactions, substitutions and restorations. *Ocean and Coastal Management, Vol. 46*, p. 27-46.

National Geographic society (1990) The emerald realm: Earth's precious rain forest. NGS, Special Publication Division, USA.

Neba. S. (1982) Modern Geography of the Republic of Cameroon. Neba Publisher, Cameroon.

Ngome, M. (1992) Environmental education in Cameroon: Problems and Prospects. WWF, Cameroon, Yaounde.

ONADEF (1993) Cameroon's forests: for a sustainable and lasting management. National Forestry Development Agency, Yaounde.

Pearce, D. W. (1994) Valuing the environment: Past Practice, future prospects. In: Ismail, S. and Steer, A. (eds.) *Valuing the environment. Environmentally Sustainable Development Proceedings Series No.2* World Bank. Washington, D.C.

WWF (2003) A users manual for building resistance and resilience to climate change in natural systems. World Wildlife Fund, New York, 240 p.

Chapter Eleven

Climate Change and Management of Lake Basin Ecosystems

Summary

Lake basins provide valuable direct goods and services to local communities and an enormous array of plants, animals and microorganisms depend on the lake basin ecosystem for their survival. Moreover, these basins continuously channel rainfall from the surrounding landscape through the interconnected lakes, rivers and wetlands that lie in lower areas. They can therefore be very sensitive to human-related stresses and increasing climate variation due to global warming. This chapter reviews the lake basin resources of Cameroon, identifies the underlying physical processes that are most vulnerable to change, and determines how the changes in these physical features might affect local biodiversity. It concludes that lake basin ecosystems are not only sensitive to water temperature, volume and flow rates, but are also expected to display a wide variety of changes in response to global climate change. Finally, the chapter provides data and strategies for designing and managing lake ecosystems in Cameroon in the context of human stresses and global warming.

Key Words: Climate change, lake basin, lake ecosystem, physical features, physical and biological stresses, resistance, resilience, sustainable management.

Introduction

The Cameroon volcanic line or the Cameroon volcanic province lies between latitude 4^0N and 8^0N and longitude 8^0E and 16^0E, covering the region from the Gulf of Guinea in the West through Fomban, Mayo Darle to the territorial boundary between the Central African Republic as well as the Southern border of the Chad lowlands. The area is characterized by a series of alternating rift basins (grabens) and mountain blocks (horsts) aligned in a SW – NE direction. Volcanic activity in the last 100 million years has led to the formation of a chain of mountains of sufficient height and extent for the development of one of the rarest of African habitats, that of montane forests and Crater Lake basin systems. These lakes and their surrounding montane forests provide vital ecosystem services, have cultural values and support local livelihoods. In addition to humans, an enormous array of plants, animals and microorganisms depend on the lake basin ecosystems for their survival. Because the lake basins continuously channel rainfall from the surrounding landscape through the interconnected lakes, rivers and wetlands that lie below, they can be surprisingly sensitive to intensive land alteration by human-related activities and increasing climate variation due to global warming.

The chapter reviews the lake basin resource potentials of Cameroon, identifies the underlying physical processes that are most vulnerable to change, and determines how changes in these physical features might affect the resident flora and fauna. Because lake ecosystems are sensitive not only to water temperature, volume and flow, they are expected to display a wide variety of changes in response to global climate change. The purpose of this chapter is to provide data and strategies for designing and managing lake ecosystems in the face of climate change.

Lake Basins of Cameroon

There are three main types of lakes in Cameroon. The classification is based on their mode of formation. Depression or tectonic lakes are formed by the down warping of the surface of the earth to produce depressions, which are then filled with stagnant water. These include Lake Ejagam west of Mamfe, Lake Ossa west of Edea and Lake Chad.

Volcanic lakes, on the other hand, result from volcanic activity and here water occupies hollows or craters of extinct volcanic regions and are often called crater lakes. In Cameroon, they lie along the Cameroon Volcanic Line, which is a region occupied by geologically young volcanic rocks. This Line runs from the island of Pagula in the South Atlantic, through Bioco, onto Mount Cameroon and includes the Rumpi Hills and the Manengouba Mountains. Further north it bifurcates around the Bamboutous Mountains into two arms, one running northwards into Nigeria to Biu Plateau and the other arm north eastwards to the Ngaoundere plateau.

A total of 43 volcanic lakes have been documented within the Cameroonian sector of this Line, two on Bioco and one on Pagula. The largest of these volcanic lakes is Lake Barombi Mbo in Kumba. The other main ones include Lakes Nyos, Monoum, Manengouba, Oku, Awing and Tison (Table 1)

Artificial or man-made lakes are those formed by building dams across rivers or valleys in order to hold back large quantities of water. The main ones include the Lakes Bamenjin on the River Nun, and the Mbakaou on the River Djerem. The Edea dam on the River Sanaga, the Lagdo dam on the river Benue South of Garoua, and the Mefou dam on the river Mefou East of Yaounde, are other examples. Smaller man-made lakes also exist such as those in Yaounde where according to, aerial photographs of the town taken in 1951 show wide valleys without lakes but today, large pools of water are seen in these valleys.

Lagoons constitute another category of lakes in Cameroon. They are produced by the accumulation of sand and silt to block portions of water as seen in the coastal basins, in the vast plains of the Logone River as well as in other hinterland plains. Lake Chad is a remnant of an inland sea. Today it covers an area of 24.000 thousand km^2. It is shared by Chad, Niger, Nigeria and Cameroon. The existence of this lake depends almost entirely on the Rivers Logone and Chari which flow northwards from Cameroon and from Chad respectively. However, the capture of water from these rivers for use in irrigation purposes, the evaporation of the lake water as a result of high temperatures over the last decades and the southward advance of the Sahara desert which brings along huge sand dunes into the lake, all pose an environmental threat to the future existence of this lake.

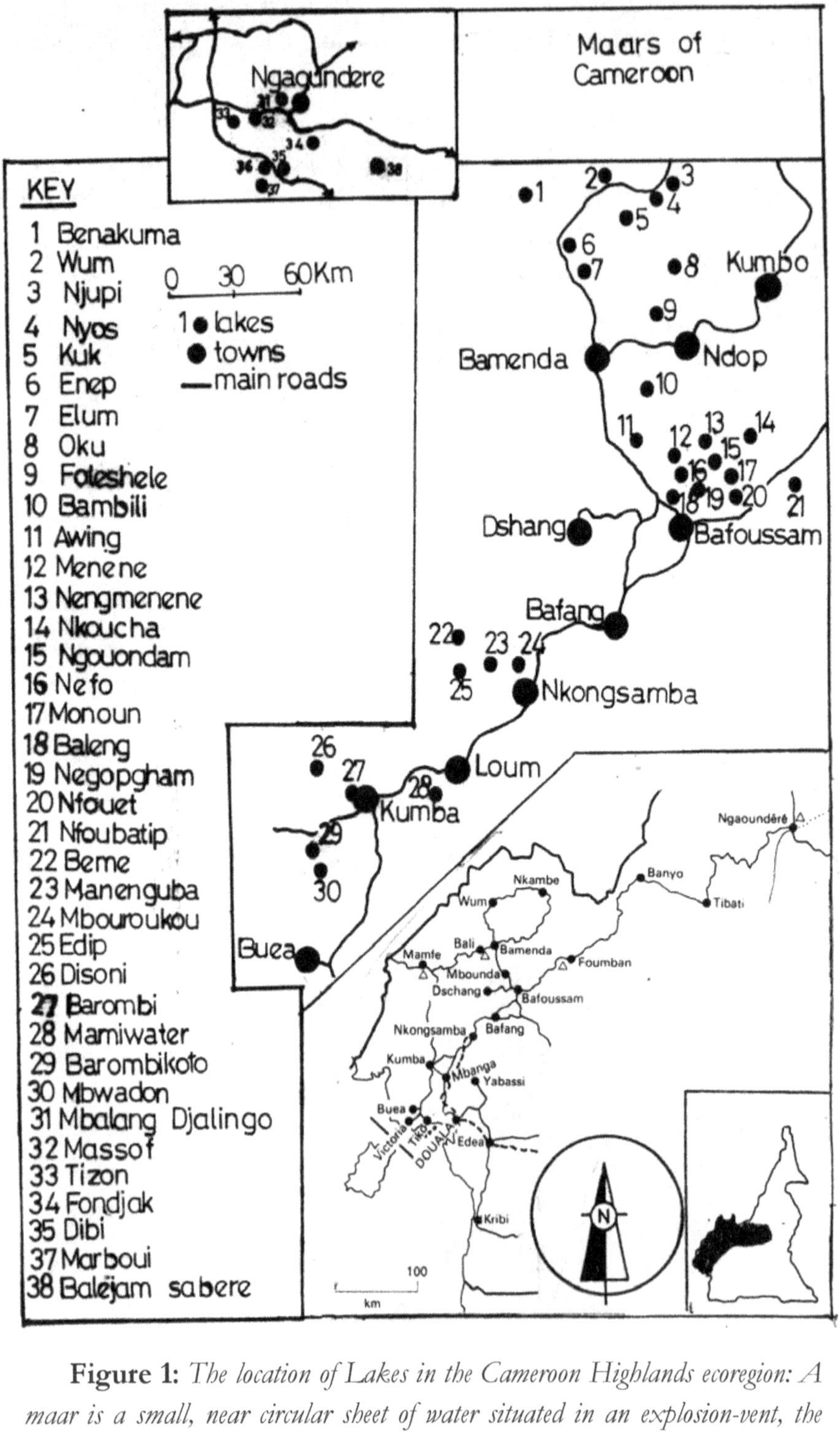

Figure 1: *The location of Lakes in the Cameroon Highlands ecoregion: A maar is a small, near circular sheet of water situated in an explosion-vent, the*

result of an eruption which has blown a hole in the surface rocks, surrounded by a low crater-ring of fragments of the country rock, but accompanied by no extrusion of igneous rocks.

Table 1: Inventory of volcanic lakes along the Cameroon Volcanic Province and their geographical locations.

Name of Lake	Latitude N	Longitude E
I. B. Kotto	$4^{0}28'$	$9^{0}16'$
Baledjam	$7^{0}08'$	$13^{0}52'$
Baleng	$5^{0}33'$	$10^{0}26$
Bambili	$5^{0}56'$	$10^{0}15'$
Bambuluwe	$5^{0}52'$	$10^{0}12'$
Banefo	$5^{0}35'$	$10^{0}27'$
Baromvi M.	$4^{0}40'$	$9^{0}24'$
Beme	$5^{0}09'$	$9^{0}38'$
Bini	$7^{0}26'$	$13^{0}33'$
Dang	$7^{0}26'$	$13^{0}33'$
Debundisha	$4^{0}06'$	$8^{0}59'$
Debundsha	$4^{0}06'$	$8^{0}59'$
Disoni	$4^{0}44'$	$9^{0}17'$
Ejagham	$5^{0}45'$	$8^{0}59'$
Elum	$6^{0}20'$	$10^{0}02'$
Enep	$6^{0}18'$	$10^{0}02'$
Gegouba	$7^{0}07'$	$13^{0}42'$
L. Monoun	$5^{0}37'$	$10^{0}34'$
Maneng – F	$5^{0}02'$	$9^{0}50'$
Maneng – M	$5^{0}02'$	$9^{0}50'$
Massot	$7^{0}17'$	$13^{0}41'$
Mbalang	$7^{0}19'$	$13^{0}44'$
Mboandong	$4^{0}27'$	$9^{0}16'$
Mfou	$5^{0}33'$	$10^{0}42'$
Mfouet	$5^{0}32'$	$10^{0}35'$
Mmonoun	$5^{0}35'$	$10^{0}35'$
Nchout	$5^{0}4'$	$10^{0}32'$
Negop gh.	$5^{0}33'$	$10^{0}34'$
Ngaoundaba	$7^{0}08'$	$13^{0}42'$

Njupi	6^{0}27’	10^{0}19’
Nwi	6^{0}25’	10^{0}12’
Nyos	6^{0}27’	10^{0}18’
Oku	6^{0}12’	10^{0}27’
Ossa	3^{0}47’	10^{0}01’
Petponoun	5^{0}38’	10^{0}38’
Tizong	7^{0}15’	13^{0}35’
Wum	6^{0}25’	10^{0}03’

Source: Ayonghe and Titanji, 2001.

Figure 1 presents the distribution of these lakes along the Cameroon volcanic line. These lakes are presented in table 2 in terms of their altitudes on the highlands, surface area and depths in order to establish their sizes. The marked with an asterisk (*) indicate maars.

Table 2: Size characteristics of lakes in the Cameroon Highlands ecoregion.

Lakes	**Location**	**Altitude (m)**	**Area (ha)**	**Depth (m)**
* Barombi Koto	Mt. Cameroon Region	106	140	5.5
* Barombi Mbo		306	415	110
Debundscha		54	6	13.5
* S. Debunscha		-	-	-
* Disoni		465	165	80
* Edip		1280	2	12.5
* Mamy Wata		-	-	-
* Mbwandong		143	-	20
Baleng (Neupan)	Bamileke and Banoun Plateau	1374	8	52
Bamefo Nefo		1100	6	64
Beme		500	60	14.5
Ejagham		100	70	17
* Manengouba (f) (femme)		1920	22	168
* Manengouba (h) (homme)		1900	2	92
		-	-	-
Manengouba (e) (enfant)				
		-	-	-
Mbouroukou (dried up)		1350	8	58

Menene		1120	7	14
* Mfou (Mbapit)		1100	25	96
Mfouet		1140	13	16
* Monoun		1120	8	23
Nkougham		1080	53	-
* Negophang		-	-	-
* Ngouondam		-	-	-
Nkeung Menene		-	-	-
Nkoupcha		1120	30	12
Petponoun I				
Petponoun II				
Bambili	Bamenda	2264	28	4
* Bambuluwe	Highlands	2053	28	58
* Benakuma		576	154	138
Bwenga		-	-	-
* Ekum		960	50	35
* Enop		697	50	78
Foleshele		-	-	-
* Lwi (Nyos)		1091	158	208
Mbi Crater (Foleshele)		-	-	-
Njupi		1020	30	-
* Nyi (Kuk, Ndu)		1316	50	47
* Oku		2227	243	52
Vindu Crater		-	-	-
* Wum		1177	45	124
* Balejam (Sabere)	Adamawa Plateau	1249	25	13
Bini		1079	20	-
Dang		1079	80	-
Dibi		-	-	-
Fondjak		-	-	-
* (Marboui) Gagouba		1180	20	104
* Massot		1054	5	
* Mbalang (Djalingo)		1130	50	52
Mgaoundaba (Tabere)		1160	10	62
* Tizong		1160	8	48
(maar de Mardja)		-	-	-

* Youkou (dried-up)		-	-	-

Source: Giresse et al., 1994

Photo: *Lake Tison: a crater lake near Ngaoundere.*

Crater Lakes: These are perched at the summits of strombolcan volcanic cones. In the Bamoum-Bamileke country are Lake Baleng, Nefo, Nfouet and Negopchang. Crater lakes are also found in the Wum Plateau. These include Lake Nyi, Enep, Elum, Benakuma and Lake Nyos. In the Adamawa plateau are Lake Baledjam and Balam. Violent explosions in the Kumba area opened up craters occupied by Lake Barombi, Dissoni and Mbo. Other crater lakes are Barombi, Dissoni and Mbo. Similar explosive lakes are found in the Bamenda area: Lake Oku, lake Foleshele, lake Bambili and Awing. Lake Bambili and lake Awing are probably calderas with large, deep craters. These probably collapsed and coalesced at an altitude of 2300m and therefore appear as twin lakes. In the western Highlands lake Mfou is found on the summit of Mount Mbapit. The lake has a diameter of 100 metres. It occupies the bottom of a volcanic vent surrounded by rhyolitic vertical walls that rise above the water level by over 200 metres. On the Manengouba mountain are two lakes: The "Woman lake" has a surface area of about $176{,}400m^2$ and has steep slopes of about 50^0 that rise 165m above the water surface. The "Man lake" has a surface area of $84{,}00m^2$.

Volcanic – Dam lakes: The course of the river Nkam which drains the Mbo plain has been dammed by recent basaltic flows that issued from the Mount Manengouba. A lake developed behind this barrier. The Mbam cut through the barrier of basalt rocks to form a series of rapids, gorges and waterfalls downstream of Kekem. These gorges remain today. The Ekom fall which has a plunge of 80 metres is one of these falls. Late Quaternary climate fluctuations probably accelerated these processes.

The Ndop plain upstream of Bamendjin had a similar barrier. Deposition on the Ndop plain is a consequence of the Bamendjin ridge. A lava flow from a small volcanic phase in the region blocked the river Noun which drains the plain at the lower part. This volcanic ridge dammed the river to create a lake which gradually experienced sedimentation. This is the reason for the alluvial morphology of the Ndop plain. Late Quaternary climate fluctuation and anthropic transformation of space accelerated this process.

On the Bamoum plateau are small lakes such as the Petponoun and Monoun. These are a consequence of the disorganization of the original drainage network by lava and cinder projections that crossed river valleys. These projections resulted from explosive volcanic activity.

In the central Adamawa basalt projections totally disorganized the flow of streams. This created zones of alternating lakes and swamps depending on the season. These can be seen north of Ngaoundere. These are either swamps or lakes.

The lakes between Benue and Hossere Sidiri in the Benue Basin are probably formed in this way. One also finds a similar process in the Menchum valley which has been blocked by basalt flow around Befang to create small lakes. Lake Njupi in Nyos (Wum) is a volcanic-dam lake.

Flood plain Lakes: These are lakes whose origin is not related to volcanic activity. They are lakes formed on alluvial plains. These are either located along rivers with large valleys and meanders or along river valleys whose water inputs from runoff, rainfall, and ground water are seasonal such as the Benue. They are also located at the mouth of large rivers on the low coastal plain such as the Sanaga and Nyong Rivers.

The Benue, from the confluence of the Mayo Godi to Faro, and the Mayo Kebi flows in a large alluvial valley that is braided by its load and meanders. The sediments have built levee or raised banks along the Benue with a high dry terrace and a floodplain terrace. On the flood plain are developed water masses or lakes that are more or less permanent. Some occupy cut meanders. These are ox-bow lakes. Some occupy local depression on the flood plain. The Benue flows in levees or raised banks and its tributaries find it difficult to join it. The levees make the tributaries almost parallel to the river for some distance. This causes marshy areas and small bodies of water as seen west of Hossere Sidiri. Another example is the Vinde Deuloumi on the Mayo Tchikali.

For the same reasons the confluence of the Faro and Benue is a flooded marshy zone during the rainy season. A series of small lakes are present during the wet season.

On the upstream side of the Sanaga delta, at the hinterland of the mangrove swamps are Lake Ossa, Tisongo, and Mboli. Lake Ossa has a digited shape and small islands. It resulted from the flooding of a small tectonic depression probably between the Cretaceous and Paleocene. In the neighbourhood of Lake Ossa is Lake Mevia. It lies east of Lake Ossa. Lake Mevi occupies a fault line valley. The steeply scarped western bank of the lake is evidence of tectonic forces. Lake Mboli, Ve and Tisongo were probably formed in the same way.

Observed Climate Change Impacts on Lake Basins

Severe Climatic Conditions: Lake Chad is shared by Chad, Cameroon, Niger and Nigeria. South of the lake seasonal swamps result from flooding by the Logone and chari River system. The depth of the lake varies yearly and seasonally. In the north-western portion it is usually from 4 to 7m with 10m maximum in certain areas of the archipelago. In the Southeast, alluviation by the Chari River reduces the depth which averages about 3 to 4m near the Chari entrance. Along the archipelago the depth is 10m. The Lake Chad is related to depth by (Major *et al.*, 1974):

$$Vol = 2.4 \times 10^9 (h - 278)^2$$

Where: Vol = Volume in m^3

h = depth on IGN reference (4.13m at Bol corresponds to 282m IGN).

Concerning seasonal fluctuation, the probability is 50% that maximum height will be reached between 25 October and 8 November, minimum between 20 April and 20 May. Over the past 20 years or so, the annual rise and fall have varied from 13m to 1.82m and from 0.6m to 1.22m respectively. The rise depends mainly on the magnitude of the Chari flood, and the fall of evaporation between January and July.

The lake receives inflows from Surface tributaries and rainfall. Average annual inflows from the tributaries are:

Chari	40.4×10^9 m^3
El Beid	2.1×10^9 m^3
Komadugu Yobe	0.5×10^9 m^3
Yedseram	0.1×10^9 m^3

The average annual rainfall on the lake is 351mm. During the wet period, the average lake surface area is about 11.000 km^2. Because of surface area fluctuations, however, the average influx from rainfall is 3.33×10^9 m^3. Total yearly inflow, therefore totals 46.3×10^9 m^3. Losses from the lake occur by evaporation (90%) and by infiltration (10%) into the banks of the islands in October and November, with some returns to the lake during March and July when lake level is low. Average annual loss by evaporation and infiltration is about 2.3m. The lake Chad Basin Commission (1972) also reports a natural spillage or overflow from the lake at Bohr-el-Ghazal, which drains northward to the Bodele depressions. The size of this lake may in future be reduced drastically if the process of river capture now taking place in the northeast of Garoua is completed. It will have been completed when the head waters of the River Logone have been diverted into the River Kebi, a tributary of the River Benue.

The effects of climatic changes on landform are clear in the Chad basin. Climatic conditions all over Africa have varied considerably during the Quaternary period of the last million years. The changes can be deciphered in the soils and present day drainage systems. Lake Chad for example which is less than 8.5m deep for the most part, and was reduced to a little more than a swamp early in this century, was about the size of the Caspian Sea, possibly as late as 5000 or 10,000 years ago. The old shoreline can be traced across Bornu in Nigeria and far to the northeast. The lake must have been over 165m deep, its water overflowed into the Benue Valley to reach the Atlantic (Grove, 1970).

Siltation Processes: Some lakes are gradually being filled up with silt brought down by rivers flowing from mountainous regions, for example, Derwent Water and Bassenthwaite in the Lake District of England. In the Western Highlands of Cameroon lakes have dried up in this way. Oral tradition mentions the migration of lakes in Bambili, Awing and Befang; where the old lake basins are still visible. In the Bamileke Plateau lake Mbouroukou dried up in this way. The late Holocene-Pleistocene pluvial-arid climatic oscillation, anthropic transformation of space and current climate variability has certainly contributed to these lake basin changes. In recent years the reduction in the depth of water in lakes of the Bamenda Highlands has called for concern. Most conspicuous is Lake Awing which is a reservoir for the water supply of Santa. During the dry season the lake outlet has been observed to let out only extremely low water discharges.

Observed Non-Climate Impacts on Lake Basins

The lake basins with their healthy montane climates combined with their rich volcanic soils attract both animal and human activity. An enterprising agriculture has developed around most lakes. The short montane grasslands around them also attract pastoral activities. Some of these lakes are surrounded by sub-montane and montane forests which are the climatic climax formations. The present day grasslands are the result of decades of forest degradation by slash-and-burn shifting cultivation and the invasion of forest by cattle. These grasslands are therefore pyroclimax types.

Most of the lakes are breached craters, that is, lava was ejected to one side of the crater only. Therefore, streams flow from most of them. These serve local communities in the lower lands. Deforestation and overgrazing of lake basins threaten the sustainability of water resources. Accelerated erosion resulting from anthropic activities is responsible for the decline in the water yields of lake basins.

Photo: *Cattle and sheep congregate in a lake shore in Adamawa plateau causing intense soil erosion in the area. Degraded Sudano-Guinean Savanna is seen at the background*

The degradation of Crater Lake environments is seen against the background of their being under pressure because of their fertile soils and diverse biological resources. Due to this, animals and man contaminate the lake waters and environment. Human activities around the lake constitute the principal cause of contamination by man and animals. While man's actions particularly defecation, bathing, the use of fertilizers on farms on the slopes surrounding the lakes and the deposition of toxic substance (dry cell batteries) remain the major sources of contamination, the grazing and watering of cattle around some of the lakes also contribute to their contamination. The ignorance of the importance of the environment and biodiversity makes them over exploit and overgraze the plant life, fertile soils and pastures around the zone, thereby reducing the biodiversity. This could also be as a result of the low incomes of population as well as the lack of alternative income sources. Actions would be needed to protect the biodiversity of the lake areas. There is an urgent need to clarify human non-activity areas around the lakes while setting up sensitization campaigns to raise awareness about the dangers of the disappearance of the rich plant and animal species around the lakes. This can not easily be done without adequately involving the local population. The issue of the use of an appropriate participatory approach thus becomes imperative.

The current threats to the lake basins can be summarized as follows:

- **Water diversion and withdrawals**: In order to meet the agricultural needs of a growing population, large quantities are withdrawn directly from some lakes or diverted from rivers flowing into lakes. This is the case of the Lake Chad. Market gardening in the Western Highland in montane environments also depends on such withdrawals.
- **Nutrient enrichment**: Runoff from fertilizers used in agriculture adds large amounts of nitrogen and phosphorus to lake waters. The added nutrients lead to the excess growth of algae (which is sometimes toxic) as is the case in Lake Chad. This results in reduced water clarity and light penetration. Because of this increased primary productivity, the activity of decomposing, oxygen-consuming bacteria increase and oxygen levels decline. Shifts in the food web and alterations in bottom-water habitat can lead to changes in species composition and distribution.
- **Toxic pollution**: Most of the toxic pollution is derived from agricultural activities within and around lake basins. There is the indiscriminate use of pesticides in these areas despite the fact that these sites also serve as natural water reservoirs for local communities. Because many pollutants accumulate in fatty tissues, they can magnify in food chains and can reach harmful concentrations.

Destructive land use practices: These are practices that result in vegetation loss within lake basins. The forests and native plant communities surrounding lakes help protect water quality and quantity by filtering and sorting runoff. Changes in land use brought about by agriculture and grazing lead to increased runoff with higher levels of nutrients and other pollutants. In addition more sediments are washed into the lake reducing its capacity while raising the turbidity. This has negative impacts on fish, filter feeders, aquatic plants, and bacteria (Baron *et al.*, 2003; Megahan *et al.*, 1992). See figures 2 and 3.

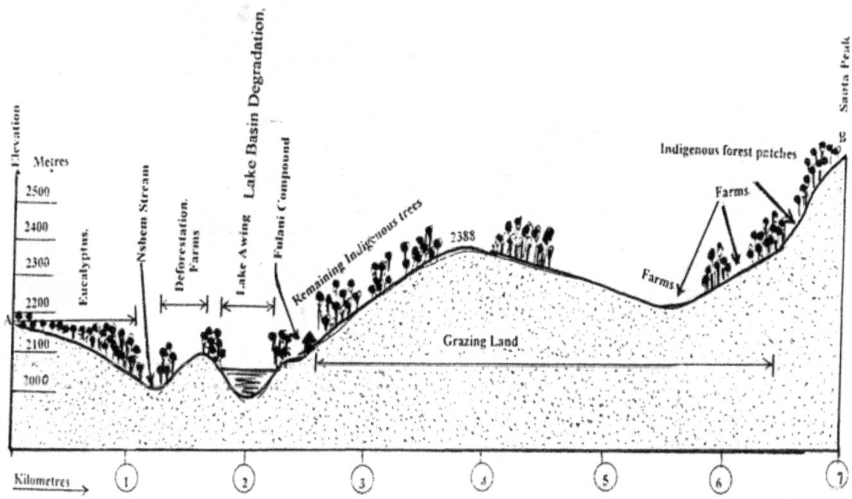

Figure 2: *Longitudinal Profile through Nshem stream – Lake Awing and Mount Lefo showing critically Degraded Montane forest areas.*

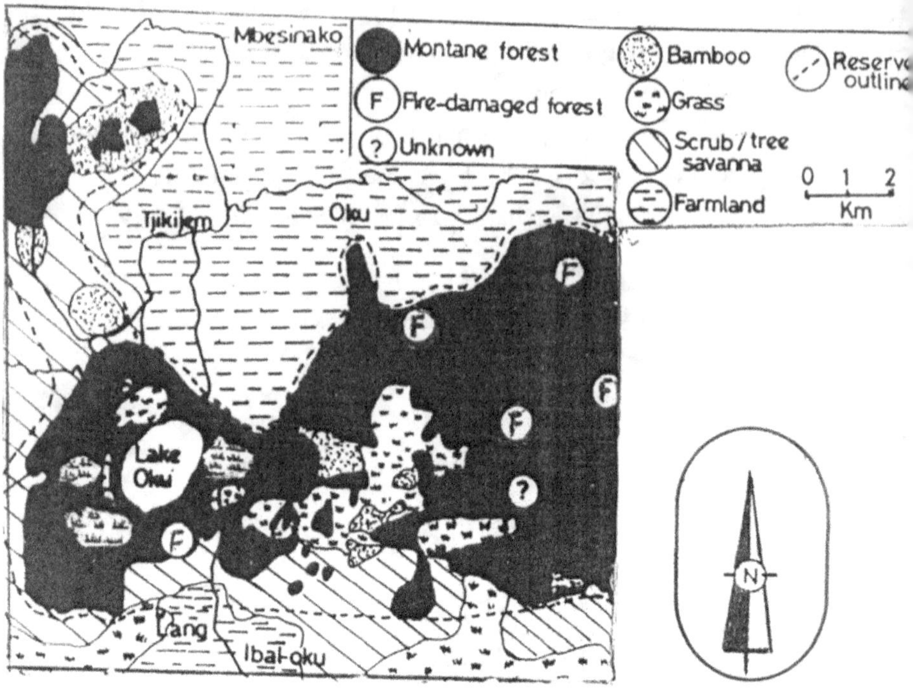

Figure 3: *Human-related pressures in the Lake Oku Basin.*

Physical Effects of Climate Change on Lakes

Increased mean surface temperatures and evaporation in many lakes (Schindler, 2001; Zinyowera *et al.*, 1998) is anticipated. If precipitation does not increase enough to compensate, this could lead to reductions in outflow and/or lake volume. Important spawning or rearing habitat near the edges of lakes (Tyedmers and Ward, 2001) would be lost if lake levels declined, and lake characteristics due to water outflow could change dramatically. Lakes that currently supply outflow to downstream systems may become endorheic (with no outflow), and endorheic freshwater lakes may become saline (Schindler, 2001). Observations in the country show that these lakes are particularly sensitive to climate effects on outflow as current outflows are small, and even minor declines in precipitation are expected to completely close these basins (Bootsma and Hecky, 1993).

Changes in mean air temperature have been shown to increase water temperatures, resulting in a range of physical and biological effects. For example, studies in Lake Tanganyika, a deep tropical lake in East Africa show that water temperatures have risen by 0.2^{0}C at the lake bottom and 0.9^{0}C at 100 metres since 1913 (Verberg *et al.*, 2003). Although the resulting change in temperature is relatively small, the vertical gradient in water density (which depends on temperature) has tripled. This sharpened density gradient has reduced annual mixing, which normally supplies nutrients in the upper layers of the lake has led to a 70% reduction in primary productivity since 1975 and an increase in water clarity and light penetration (Verberg *et al.*, 2003).

It has been established that climate change may affect lake water chemistry in a number of ways. Droughts and decreased groundwater flow or groundwater recharge may make some lakes more susceptible to acidification, as groundwater often contains acid-neutralizing chemicals important to lake buffering (Schindler, 2001). However, the overall pH and chemical balance of lakes may be affected by temperature and rainfall changes in ways that are site specific (depending on the geology of the lake basin, vegetation and land use).

Biological Effects of Climate Change on Lakes

The majority of lakes in the country are shallow (Table 2). Shallow lakes with adequate amounts of phosphorus, nitrogen, molybdenum, cobalt and carbon, are very productive. Productivity also depends on the magnitude of flow of water through a lake, which is dependent on the volume of the lake and the size of its catchment area. The rate of flow of water will be greater in a small shallow lake with large catchment area and with high rainfall. Rainfall in the Cameroon volcanic province is between 2500mm to 1500mm per year. The lakes are mostly shallow with small basin sizes. A good index of the magnitude of water flowing through a lake is the replacement quotient defined by Brooks and Woodward (1956) as the volume of water contained in the lake divided by the amount of water passing through that lake daily. Lakes with a low water renewal have a high and more stable plankton community, whereas for those with a high renewal such as those in Cameroon, the converse is true. These lakes experience rapid outflows with slow and little inflow through groundwater recharge during the rainy season. During the dry season both groundwater recharge (inflow) and outflow drop considerably. These seasonal fluctuations in volume provoke fluctuations in plankton production and these are reflected in changes in the population structure of fish species.

Although the effects of climate warming on the biotic communities of tropical lakes has received little research attention, large decreases in primary productivity due to climate warming (Verberg *et al.*, 2003) are likely to have a significant impact on the rest of the food chain. Changes in lake water chemistry are also expected to have a range of effects on biological communities. Livelihoods based on lake fishing will be disrupted.

A few species of frogs are only found in the main peaks and lake basins of the Bamenda Highlands and nearby mountains where they are confined to montane zones (Stuart, 1986). The frogs which appear to be restricted to these areas (Mount Oku and Lake Oku, Mount Manenguba and the Manenguba Lakes, Bamboutos Mountains and its Lakes, Obudu Plateau in Nigeria and Tchabal Mbabo and the Adamawa Plateau Lakes include 14 species. These are endemic to the ecoregion. The *xenopus* sp.; which has become extinct in Lake Victoria due to the introduction of *Tilapia* is probably endemic to Lake Oku (Stuart, 1986) . The species of montane frogs occurring in the Oku area can be divided into two categories. Those, such as *Leptopelis nordequatorialis*, can tolerate deforestation and may even thrive on degraded grasslands around the basin. Others, such as *Werneria bambutensis* and *Cardioglossa oreas* are threatened by further destruction of their forest environment around the lakes. Climate change resulting in montane forest retreat will certainly cause the extinction of these narrow endemics.

Managing Lake Basins to Withstand Climate Change

- **Preservation of Lake Basin Biodiversity**: In aquatic systems, high biodiversity is often found in isolated habitats such as lake basins in upland and mountain areas. Such areas also harbour rare species, such as endemic species that have evolved in and remain restricted in the montane habitats. In protecting some of these rare biodiversity sites, rare or vulnerable species may also be protected. Strategies to protect the endemics of the lake basins must be accompanied by goals of protecting ecosystem functions (Junk, 2002) and increasing resistance and resilience to climate change. There is a need to protect physical features rather than solely individual species.

- **Protection of the Physical Setting of Lake Basins**: It should be noted that aquatic ecosystems are different from many other ecosystems because they are usually governed by "bottom up" rather than "top-down" dynamics – in other words, much of ecosystem function is determined by basic physical features such as the volume and rate of water flow, lake basin morphology and nutrient balance, rather than by species assemblages (Moss, 2000). Protecting flow patterns, water quality and water quantity will go a long way towards protecting biodiversity in lake basin habitats (Abel *et al.*, 2002), whereas conservation efforts that focus solely on protecting particular species or groups of species are doomed to failure. In many cases, the function of a species in a shallow lake ecosystem is actually more important than its identity; for example, plants are essential elements in shallow lakes, but the exact species of plant may be less important than the physical features it provides (Moss, 2000). Physical features of lakes are expected to undergo a number of changes as a result of climate warming and rainfall variability. Maintaining healthy, forested lake basins, and reducing the input of toxic substances will increase the likelihood that lake basin ecosystems will be able to adjust to climate change.
- **Protection of Lake Basins from Human Pressures**: Species should be protected from outside stresses in the face of local climate variability. This is important because stressed systems display reduced resistance and resilience to change (Noss, 2001). Human stresses, such as overexploitation and poor land use practices, should be reduced as much as possible.
- **Prevention of Access to Invading Species**: It is also critical to prevent access to invading species, and to eliminate or control harmful non-native species already present. The story of the invasion of Lake Victoria by the water hycinth is well known. In the same lake the introduction of *Tilapia sp.* is known to have disrupted vital food chains resulting in the extinction of the *xenopus sp.* Many systems are likely to become more vulnerable to invasions, as thermal barriers that previously excluded invaders will become removed, and communities that are already stressed by climate change are invaded by warmer-adapted species (Carpenter *et al.*, 1992).

- **Management of the Entire Lake Basin**: The lake basin ecosystems are so intricately connected to one another and to the terrestrial systems that surround the lake. Managing the entire basin rather than protecting aquatic elements or habitat patches within the basin therefore becomes very important as the effects of climate change intensify and are magnified by interactions with human stresses (Schulze, 2000). This requires the introduction of Integrated Lake Basin Management in order to buffer systems against climate change and provide a basis for long-term conservation.

Conclusion

Cameroon is endowed with several Lake Basin resources. These constitute important natural water reservoirs; constitute habitats for endemic fish and frog species, and vascular plant species. They also provide scenic sites and panoramas for tourism and recreation. The fertile volcanic soils have attracted an enterprising agriculture in these sites. The montane grasslands and healthy climate free of trypanosomiasis have equally attracted intensive pastoral activities in these afro-alpine sites. Unfortunately, there is neither a national policy for reducing human pressures on these sites nor one for building resistance and resilience to climate change. If Cameroon has to guarantee the sustainability of these lake basin resources adaptive management programmes will be imperative. Long-term protection of lake basins, adaptation strategies to climate change and sustainable land management must be enacted on a national scale.

References

Abell, R.; Thieme, M.; Dinerstein, E.; and Olson, D. (2002) A source book for conducting biological experiments and developing biodiversity visions for Ecoregion Conservation. *Freshwater Ecoregions, Vol. 2*, WWF, Washington D. C. 201p.

Ayonghe, S. and Titanji, V. P. K. (2001) Lake Nyos and other explosive volcanic lakes in Cameroon: Facts, media coverage and Myths. J. Dunlop and W. Roy (eds). *Culture and environment.* University of Stratchclyde in Glasgow/University of Buea, p. 208 – 222.

Baron, J. S.; Poff, N.; Angermeier, P.; Dahm, C.; Gleick, P. *et al* (2003) Sustaining healthy freshwater ecosystems. *Issues in Ecology, Vol. 10,* p. 1 – 16

Bootsma, H. A., and Hecky, R. F. (1993) Conservation of the African Great Lakes: Limnological perspective. *Conservation Biology, Vol. 7*, p. 644 – 656

Brooks, A. and Woodward, W. (1956) Some observations on the effects of water inflow and outflow on the plankton of small lakes. *Journal of Animal Ecology, Vol. 25*, p. 25 – 35

Carpenter, S.; Fisher, S.; Grimm, N.; and Kitchel, J. (1992) Global change and freshwater ecosystems. *Annual Review of Ecology and Systematics, Vol. 23,* p. 119 – 139

Giresse, P.; Nyos, S.; Poucet, M. (1994) Procéssus sedimentaires seculaire et geochronlogie au 210 Pb des principaux lacs de la dorsale Camerounaise. *Bull. Soc. Geol. France, Tome 165, no. 4*, p. 363 – 380

Grove, A. T. (1970) Africa South of the Sahara. Oxford University Press, Oxford. 280p.

Junk, W. J. (2003) Long-term environmental trends and the future of tropical wetlands. *Environmental Conservation, Vol. 29*, p. 414 – 435

LCBC (1972) Survey of the water resources of the Chad Basin for development purposes: Surface water resources in the Lake Chad Basin. UNDP/FAO, Rome

Major, C. D.; Kirshen, P. H. and Lengyel, Z. (1974) A framework for evaluating long-term strategies for the development of the Sahel Region: Annex 8 – An approach to water resource planning. MIT, Massachusetts, 282p.

Megahan, W.; Potyondy, J. and Seyedbagheri, K. (1992) Best management practices and cumulative effects from sedimentation in the South Fork

River: an Idaho case study. In: R. J. Naiman (ed). *Watershed management: balancing sustainability and environmental change.* Springer – Verlag, New York, p. 401 – 414

Moss, B. (2000) Biodiversity in freshwaters – issues of species preservation or systems functioning? *Environmental Conservation, Vol. 7,* p.1 – 4

Noss, R. (2001) Beyond Kyoto: Forest management in a time of rapid climate change. *Conservation Biology, Vol. 15,* p. 578 – 590

Schindler, D. W. (2001) The cumulative effects of climate change and other human stresses on the Canadian freshwaters in the new millennium. *Canadian Journal of Fisheries and Aquatic Sciences, Vol. 58,* p. 18 – 29

Schulze, R. (2000) Modeling hydrological responses to land use and climate change: a Southern African perspective. *Ambio, Vol.* 29, p. 12 - 22

Stuart, S. (1986) Conservation of Cameroon Montane Forests. Report of ICBP Cameroon Montane Forest Survey: Nov. 1983 – April 1984. Cambridge. 89p.

Tyedmers, P. and Ward, B. (2001) A review of the possible impacts of climate change on BC's freshwater fish resources and possible management responses. *Fisheries Centre Research Report, Vol. 9,* p. 1 – 12

Verberg, P.; Hecky, R.; Kling, H. (2003) Ecological consequences of a century of warming in Lake Tanganyika. *Science, Vol. 301,* p. 505 – 507

Zinyowera, M.; Jallow, B.; Maya, S.; Okoth – Ogendo, H.; and others (1998) The regional impacts of climate change: an assessment of vulnerability. In: R. Watson; M. Zinyowera; R. Moss; D. Dokken (eds). *A special Report of IPCC Working Group II,* Cambridge University Press, p. 30 – 84

Chapter Twelve

Management of Observed and Anticipated Physical Effects of Climate Change on Lake Chad

Summary

Global climate change is likely to lead to increased water temperatures and evaporation in many lakes. This certainly will have physical and biological effects on lake ecosystems. The chapter appraises the guidelines established for the management of the Lake Chad basin with a focus on the observed and anticipated effects of climate change and anthropic activities on the physical resources of the basin. It presents an overview of the basin water resources and the management lacunae. The chapter finally concludes that current management strategies do not fully accommodate observed and anticipated effects of climatic change on the availability of water resources in the lake basin. It therefore designs a strategy for the basin that can buffer the adaptations to climate change effects.

Key Words: Lake Basin, climate change, water resources, mitigation, integrated river basin management (IRBM), ecoregion conservation (ERC).

Introduction

Increased mean surface temperatures are likely to lead to increased water temperatures and evaporation in many lakes (Schindler, 2001; Zinyowera *et al.*, 1998). If rainfall does not increase enough to compensate, this could lead to reductions in outflow and/or lake volume (Tyedmess and Ward, 2001). Severe climatic conditions have caused the gradual shrinkage of some lakes, for instance, Lake Ngami in South Africa. The African Great Lakes are particularly sensitive to climatic effects on outflow, as current outflow is small, that is, only 6% of water input to Lake Tanganyika leaves as riverine outflow, and even minor declines in rainfall of 10 to 20 % are expected to completely close these basins (Bootsma and Hecky, 1993). Increasing temperatures coupled with reducing rainfall and encroaching sands from the Sahara Desert are gradually reducing the size and depth of Lake Chad (Ngwa, 1979). The size of this lake may in future be reduced more quickly if the process of river capture now taking place on the ridge separating the Mayo Kebi and River Logone is completed (Morin, 1982). Ssentogo (1979) observed that there is less ecosystem stability in temporary fluctuating lakes such as Lake Chad. It is a closed lake which is increasing shrinking because of much evaporation due to high temperatures. The increased salinity of the lake has resulted in more blue green algae production. The threat to lake Chad, which in effect is a threat to the economic life of the inhabitants of the basin is the grave concern of the countries sharing the lake, that is, Nigeria, Niger, Chad and Cameroon (Figure 1)

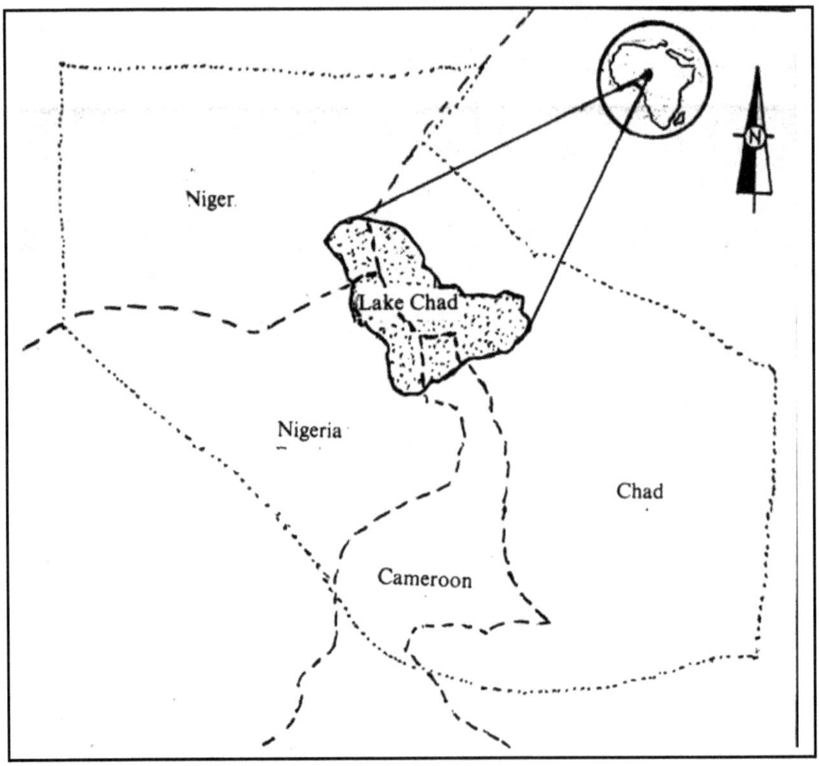

Figure 1: Location of Lake Chad and the Lake Chad Basin. Dotted lines are lake basin limits.

The paper reviews the Lake Chad basin water resources, assesses the anthropogenic threats to their sustainability under climate change conditions, the management guidelines established by member countries of the Lake Chad Basin Commission (LCBC) with a focus on the observed and anticipated effects of climate change, and identifies the management models that can mitigate climate change impacts.

Research Methods

The principal data sources were secondary. Project documents of the Lake Chad Basin Commission and other related archival material constitute the basis of the review and appraisal. These were complemented by field observations which were limited to the Cameroon portion of the basin. Aerial photographs produced by the National Geographic Institute for the study area were also used to describe changes in water dependent land use patterns. Finally, informal interviews involving a two-way conversational method were used to obtain data from local resource users and development experts working in the Cameroon portion of the basin. A limitation of the study is the lack of up-to-date data on land use changes and hydrometeorology.

The Study Area and Problem Background

Climatic conditions all over Africa have varied considerably during the Quaternary period of the last million years (Kadomura, 1986; Diamond, 1980, Hamilton, 1982; Hamilton, 1976). These changes can be deciphered in the soils and drainage systems of the present day. Lake Chad, for example, which is less than 8 metres deep for the most part and has been reduced to a little more than a swamp, was about the size of the Caspian Sea, possibly as late as 5,000 or 10,000 years ago (Grove, 1970). The old shoreline can be traced across Bornu in Nigeria and far to the north-east (Figures 2 and 3). The lake must have been over 165metres deep, its waters overflowed into the Benue valley to reach the Atlantic Ocean, and there can be no doubt that the Sudan zone at the time was much more humid then than it is now.

Conditions drier than those of the present day are indicated by deep soils derived from blown sands that cover extensive areas far beyond the present limits of the Sahara Desert, and ancient dune fields can be easily seen on aerial photographs. The deserts have evidently contracted form time to time as the climate fluctuated between arid and pluvial phases.

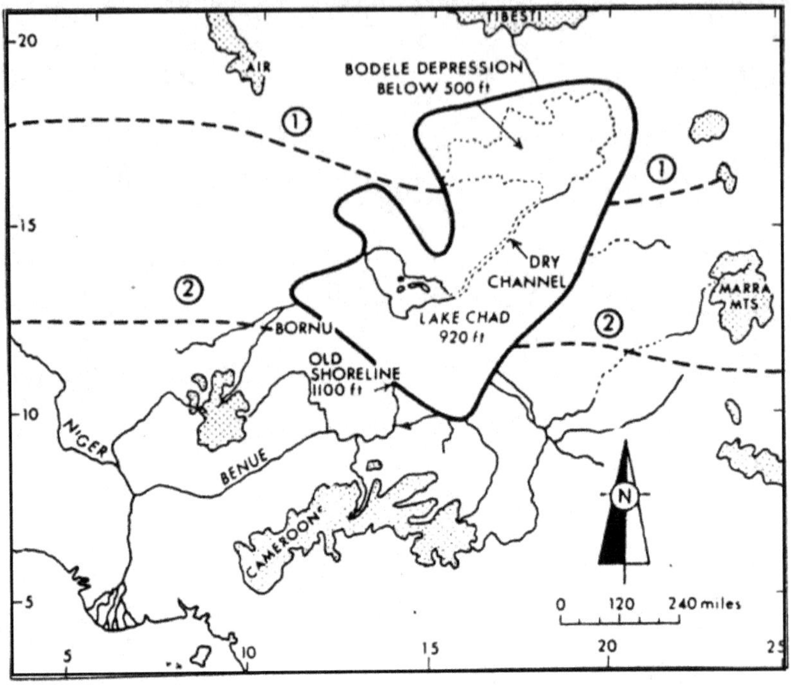

Figure 2: Shifting Lake Shores and desert limits in the Chad Basin.

The present extent of lake Chad (280m) above sea level, is compared with the old shorelines of the ancient mega Chad at (335m) above sea level (thick line), which overflowed into the Benue River. Dashed line (2) is the southern limit of old, vegetation covered dunes. It lies far south of the southern limit of moving dunes of the present day, marked by dashed lines (1). These extreme changes probably occurred in the period 20.000 – 50.000 years ago (after Grove and Warren 1968 cited by Grove, 1970).

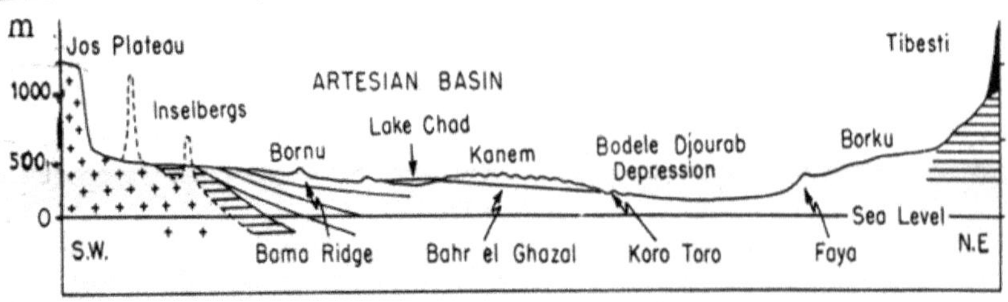

Figure 3: Cross-section of Lake Chad in a SW – NE Direction (Modified after Grove, 1970; p.7)

In the region short term climatic variations are superimposed on the long-term paleo-climatic fluctuations. Extreme variability of annual rainfall amounts is characteristic of the region (Ledger, 1964). Dresh (1973) reports that "the shorter the rainy season, the more irregular the rainfall is in space, in time over the year, and from one year to another".

The area is thus subjected to wet and dry years. Dry periods occurred from 1910 to 1914 and from 1941 to 1942. The period from 1951 to 1960 was wetter (Major *et al.*, 1974). Having statistically analysed rainfall data for the Cameroon part of the Chad Basin, Ayonghe (2001) found that high rainfall was evident from 1930 to 1950, 1968 to 1976, and 1981 to 1988. In the Chad Basin of Cameroon the data between 1960 and 1990 were in agreement with similar studies in the Chad basin in Nigeria where Carter and Alkali (1996) reported rainfall as decreasing by 8mm per year. The net change in the total amount of rainfall in Cameroon as a whole was minus 282mm between 1960 and 1990, equivalent to minus 43mm per decrease. The net change in the annual number of rainy days minus 7 days over the same period. When projected to the year 2060 there is a net rainfall decrease of 559mm and 16 days respectively.

In the same study Ayonghe (2001) working in Cameroon as a whole showed that the net trend per decade between 1930 and 2000 is a temperature increase 0.14^{0}C. In the Chad Basin the Maroua Station showed an increase per decade of 0.23^{0}C. When projected to the year 2060 there is a net increase of temperature of 1.8^{0}C for the whole of Cameroon and 1.4^{0}C for Cameroon's Chad Basin. Droughts and flood are becoming frequent.

The Lake Chad Basin Commission (LCBC) was created in 1964 has the countries sharing the lake as members. Its main objectives are to promote the development of forestry, fisheries, agriculture, animal husbandry and the sustainability of water resources in the basin, in order to improve the socio-economic life of the population using the water resources based on an Integrated River Basin Management (IRBM) approach. These include the lake and the rivers draining into it.

Lake Basin Water Resources

Lake Chad is shared by Chad, Cameroon, Niger, and Nigeria (see Figure 1). South of the lake, seasonal swamps result from flooding by the

Logone/Chari system. The depth of the lake varies yearly and seasonally. In the North West portion it is usually from 4 – 7 m with 10m maximum in certain areas of the archipelago. In the southeast, alluviation by the Chari River decreases the depth which averages about 3 – 4m near the Chari entrance. Along the archipelago the depth is 10m. Lake Chad Basin Commission (1969) reports that volume is related to depth by:

$$Vol = 2.4 \times 10^9 (h - 278)^2$$

Where:

Volume is in m^3

H = depth on IGN reference (4.13m at Bol corresponds to 282m IGN)

Photo: *Dry river bed of the Mayo Tsanaga in the Chad Basin of Cameroon (Ndenecho)*

Concerning seasonal fluctuation, the probability is 50 percent that maximum height will be reached between 25 October and 8 November, minimum between 20 April and 20 May. Over the past 20 years, the annual rise and fall have varied from .13m to 1.82m and from 0.6 to 1.22m respectively. The rise depends mainly on the magnitude of the Chari flood, and the fall of evaporation between January and July.

Photo: *The River Yobe in Bornu. It is April, near the end of the dry season, and flow has ceased. The high bank on the right is probably not far from the margin of the ancient Lake Chad.*

The lake receives inflows from surface tributaries and rainfall. Average annual inflows from the tributaries are:

Chari	$40.4 \times 10^9 m^3$
El Beid	$2.1 \times 10^9 m^3$
Komadugu Yobe	$0.5 \times 10^9 m^3$
Yedseram	$0.1 \times 10^9 m^3$

Average annual rainfall on the lake is 351mm. During the wet period, the average lake surface area is about $11{,}000 km^2$. Because of surface area fluctuations, however, the average influx from rainfall is $3.23 \times 10^9 m^3$. Total yearly inflow, therefore, equals $46.3 \times 10^9 m^3$.

Losses from the lake occur by evaporation (90 percent) and by infiltration (10 percent) into the banks of the islands in October and November, with some returns to the lake during March and July when lake level is low. Average annual loss by evaporation and infiltration is about 2.3m. LCBC (1972) reports also a natural overflow for the lake at Bahr-el-Ghazal, which drains northward to the Bodele depressions. No lake depth for beginning of overflow is given, but overflow reportedly occurred when the level was 282.97m (IGN) (5.10m at Bol). A reference level of 282m (IGN) corresponds to 4.13m at Bol. LCBC (1972) also notes that the capacity of the overflow channel is small and should not be relied upon to regulate lake levels. If the capacity were increased, however, the channel might serve this function.

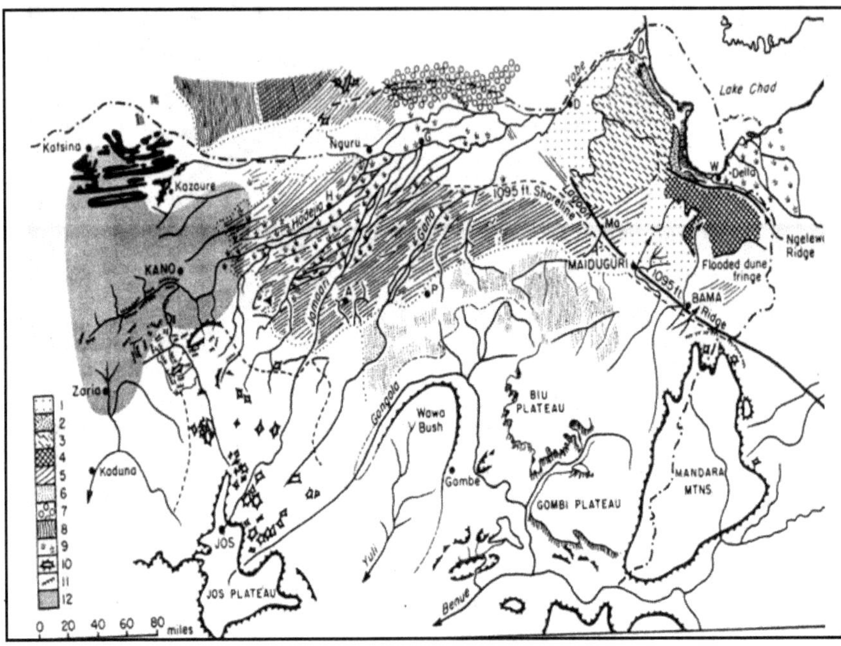

North-east Nigeria is made up of the Chad basin, floored with Pleistocene sediments, and dissected plateau country drained by the Benue River. The populous region around Kano lies on the Chad-Niger water parting. I – old deltas formed in a larger Lake Chad, 2 – sandhills bordering the present lake, 3 – NW./SE. sand ridges, with lake sediments between the ridges nearer Chad, 4 – black lagoonal cracking clays, 5 – longitudinal dunes, covered in vegetation and running ENE./WSW, 6 – open woodland showing a N./S. banded pattern, 7 – sandy plain with closed depressions, some of which contain lakelets, 8 – N./S. dune patterns, 9 – seasonal swampland, 10 – rocky hills, 11 – gullied banks of watercourses, 12 – heavily settled country with well over 200 people to the square mile. H

– Hadeija, G – Geidam, D – Damasak, Ma – Magumeri, P – Potiskum, A – Azare, W – Wulgo.

Figure 5: *The Chad Basin in North east Nigeria: Physiography and drainage (After Grove, 1970).*

The Chari system, composed primarily of the Chari and Logone rivers which join at N'Djamena, is very complicated. Since the major development area is down-stream of Miltou on the Chari and Lai on the Logone, these stations can be taken as reference points. Average annual flow near Lai is 17.7 x 10^9m^3; that at Miltou is 30.14 x 10^9m^3. The total inflow from these systems, therefore, is 47.8 x 10^9m^3. However, the total flow at N'Djamena is 40.4 x 10^9m^3 annually. There is obviously a large loss due to natural diversions when flood waters exceed river channel capacities. Part of the annual flood loss from the Logone and Chari overflows from the Logone into the Yaere flood plain, which feeds the El Beid. (The total input to the lake from El Beid is 2.1 x 10^9m^3 annually). The El Beid thus experiences two peaks, first due to local rains on the Yaere, the second because of overflow from the Logone via the Yaere flood plain. Another important diversion from the Logone is just upstream of Bongor. When the flow between Lai and Bongor reaches 2000m^3/sec, there is a small overflow to Mayo Kebbi. The Mayo Kebi runs into the Benue River, a tributary of the Niger River (Figure 5).

Yedseram: Little quantitative information is available for this system. Komadugu Yobe: This tributary, whose drainage area is 147.840km^2, enters the lake from the west end. Less than 10 percent of its total runoff reaches the lake. (Figure 4). The losses are due to infiltration and overspills. Average monthly discharges at Damasak and Yau are in Table 1.

In Toto, the rivers entering the lake bring in about 3 million tons of sediment per year and an equal amount of dissolved solids (salt and silica). The water is, however, fresh and suitable for irrigation. Researchers are still searching for an explanation of the lake's freshness.

Table 1: Average Monthly Stream Flows (Approximate) at Selected Stations in the Chad Basin (m^3/sec)

Month	Lai	N'Djamena	Damasak	Yau
January	130	1,000	30	30
February	60	500	20	10
March	70	450	20	10
April	70	420	20	10
May	90	330	0	10
June	170	420	0	0
July	500	700	20	10
August	1,080	1,440	30	30
September	1,340	2,560	40	30
October	1,560	3,500	50	30
November	520	3,400	60	30
December	210	2,000	65	40

Source: Major, 1972.

Other rivers of importance in Chad are the Bahr Sara, chiefly used for floating logs into Chad from the Central African Republic, and the Batha and Bahr-el-Ghazal, which are dry.

Multipurpose Use of Lake Basin Resources

There is local traffic on the lake itself. The Chari River is navigable between Sarh and N'Djamena from July to December. The river from N'Djamena to Lake Chad is navigable the year round and is much used for shipping. The Logone River is navigable only from N'Djamena to Moundou from August to October. ECA (no date) reports that 100,000 tons of fish are caught annually in the Chari and Logone Rivers and Lake Chad. Dekker (no date) states that programmes have been established to increase production and improve processing.

LCBC (1972) reports that flow diversion is possible from the Logone just above Bongor at Ere to Mayo Kebi and Benue River. A danger of this diversion lies in possibly upsetting the ecology of the Yaere flood plains as well as the irrigation schemes on El Beid. However, it was estimated that if flood flow in the Logone was sufficient to reach a peak of $1800m^3/sec$ at

Bongor, no danger would be done to the Yaere flood plain or El Beid irrigation. 1500m^3/sec at Bongor would insure preservation of the fisheries and pasturage on the Yaere plain; an additional 300m^3/sec is required to insure irrigation on El Beid. The average annual diversion from the Logone to Mayo Kebi and Benue River would be 0.2 x 10^{9}m^3 and would have no effect on the level of Lake Chad.

LCBC (1972) reports that a canal is possible between El Beid and Serbeouel River (just South of Lake Chad). Flood peaks occur at different times on these rivers and there is thus the opportunity to transfer flows between them so that each has a steadier regime (LCBC 1972). LCBC (1972) reports that the channel of the Bahr-el-Ghazal could be improved to help regulate the level of Lake Chad.

Photo: *Reed boats on the shore of Lake Chad used for Transporting fish, salt and passengers*

There have been only a few dams or storage sites mentioned in the LCBC reports; none are on the Chari. However, some project proposals concerning the construction of dams have been made (Table 2). It has been proposed (LCBC 1972) that a series of dams be built on the headwaters of the Logone at Koumban on the Vina River, Gore on the Pende River, and on the Logone River itself at Moundou. These dams would limit the flood flow at Lai to a maximum of 1500m^3/sec which would reduce the overspill at Lai. This would decrease water loss to Mayo Kebi and evaporation; increase the supply for Lake Chad, and make more water available for irrigation and navigation by flow release during the dry season. The dams would be of earth construction with concrete spillways and radial sluice gates. The gates would be closed in mid-August and opened gradually in October to provide flow not exceeding 1500m^3/sec at Lai. The main purpose of the Moundou dam would be storage for very large floods. The Koumban Dam would store 5.0 x 10^{9}m^3 and the Gore Dam would store 2.8 x 10^{9}m^3.

Photo: *Cattle being watered from artesian boreholes in Bornu- Nigeria (Chad Basin). This is dry season and forage is scarce.*

It has been proposed to build a dam on the Mayo Kebi at Mbourao or Dao Koumi to store water for a proposed hydroelectric plant at Chutes Gauthiot near the Chad-Cameroon border. The dam would be filled by the end of October with a volume of 10^{9}m^{3-}. This would provide enough water for a year. However, in a poor year, water would have to be withdrawn from the Logone during October (about .9 x 10^{9}m^3). If the Ere diversion were in effect, not enough flow would be provided at Bongor, and the Yaere Plain and the El Beid systems would be upset.

Table 2: Hydrographs at Diversion and Dam Sites (m^3/sec)

Month	Ere	Touboro	Gore	Moundou	Mbourao	Geidam	Gashua
January	125	38	36	100	30	30	30
February	100	38	36	100	30	20	20
March	90	38	36	100	30	20	20
April	90	38	36	100	0	0	20
May	100	0	0	0	20	20	20
June	125	47	44	125	20	30	30
July	500	150	142	400	20	40	60
August	1100	375	355	1000	30	50	110
September	1750	525	497	1400	50	50	160
October	1800	375	355	1000	80	75	200
November	750	113	107	300	60	95	120
December	250	47	44	125	50	65	40
Yearly Vol. x $10^9 m^3$	17.6	4.65	4.4	12.4	.73	.95	1.68
Mean Annual Flow	557	148	140	394	23.1	31.2	52.2

Source: LCBC, 1972

LCBC (1972) reports that the development of depression storage between Geidam and Gashua on Komadugu Yobe and between sites could be possible. 10, 388, 500m^3 of water could be stored between Geidam and the Lake. LCBC (1972) says this would not affect the ecology of the Yobe Valley; no reason is given. They report the best way to operate the system would be to fill the depressions during the flood and then provide water for three months after the flood (Figure 5). There is the possibility of storage on the Hadejia River in Nigeria. LCBC (1972) also recommends that more upstream storage sites be found on the seasonal rivers (El Beid and Komadugu Yobe).

Average monthly flows at diversions, canals, and dam sites are shown in Table 2. The source is the LCBC (1972). The values at some sites were estimated by comparing their average annual flow values with the averages at sites where the complete hydrograph was known. There is no data available on the Hadejia. LCBC (1972) describes the present and potential irrigation sites in the basin. These are listed in table 3. Most of the future potential in Nigeria and Niger has been fully developed. The inadequate supply of surface water for irrigation has resulted in the excessive extraction of groundwater reserves through boreholes in these areas. This is certainly going to upset the delicate hydrological balance sustaining Lake Chad.

Table 3: Irrigation in the Chad Basin

Location	Present	Future	Comments
CAMEROUN Yagona	3000 ha (controlled flooding) 5000 ha (seasonal)	Possible expansion	Drainage to Mayo Guerlo
CHAD Yagona Bol Dune	2500 ha (controlled flooding) 500 ha year round 4000 ha (reclaimed swamps)	Possible expansion	 Water from Phraetic wells
NIGER Diffa	275 ha (controlled flooding)	Small expansion	
NIGERIA Gamboru	1000 ha	Fully developed	Water from El

			Beid (seasonal)
Between Ngala and Lake	400 ha	16,000	Year round from lake
Komadugu Yobe (Yau, Daya, Abadan)	1600 ha	20,000	All from K. Yobe one crop in wet season
Mallan Fatori,	50 ha		From lake
Jerre Bawl	1500 ha		From Ngadda River

Source: Major et al., 1974

LCBC (1972) reports that by 2020 an area of 260,000 ha will be irrigated using water from Lake Chad sources. They estimate they will require $3.9 \times 10^9 m^3$ of water, or .5 litre/sec/ha. If we assume year round cropping at 1 litre/sec/ha, the total water need will be $7.8 \times 10^9 m^3$. It can be shown using a formula given in LCBC (1972) that this will lower the equilibrium level of the lake by 76cm, approximately one meter.

Assume the average lake level is 4.13m (at Bol) or 282m. Since the 1971 level was 280m and the lake has split apart, assume 281 is the lowest level the lake can safely have. Therefore, the year round irrigation potential from Lake Chad is 260,000 ha in an average year. The potential from the rest of the area totals about 18,000 ha, most of which is seasonal (Summary Table 3).

Water Management under Droughts

The potential for irrigation in a series of drought years is non-existent. As has occurred now, the lake has split apart and is not suitable for supplying irrigation water. In a drought, it is expected that most of the seasonal agriculture would be eliminated.

There appear to be two storage-regulation schemes that are possible in the next 20 – 30 years. They are the storage of water between Geidam and the lake on the Komadugu Yobe and the construction of headwater dams on the Vina, Pende, and Logone.

Storage at Geidam on Komadugu Yobe would provide storage for $10 \times 10^6 m^3$ of water for three months after the flood. Since less than 10 percent of the runoff from this basin presently reaches the lake, and this system provides so little of the input to the lake, this storage would have no effect on the level of the Lake and it is expected it would not upset the Komadugu Yobe's ecology. If an average depression depth of three meters is assumed, this implies the surface area covered would be $3.3. \times 10^6 m^2$. Using the evaporation rate from the lake, this implies a three month evaporation loss of :

$$\frac{3}{12} . 2.15m\ (3.3. \times 10^6 m^2) = 1.7 \times 10^6 m^3$$

Therefore, the approximate net amount available for crops would be $8 \times 10^6 m^3$.

If it is assumed that 50 percent of this reaches the Lake and is available for year round irrigation, this implies:

$$\frac{50 \, x \, 10^6 \, m^3}{365 \, x \, 24 \, x \, 3600 \, x \, 10^{-3} \, m^3 \, / \, \text{sec} \, / \, Ha} = 1600 \text{ hectares}$$

It should be noted that this storage will probably keep the peak at Bongor below $1800 m^3$/sec and prevent sufficient water from spilling over the Yaere flood plain and El Beid. Therefore, to prevent upsetting the ecology of the flood plain and the irrigation schemes on the El Beid, water will probably have to be pumped.

It is expected that under drought conditions the depressions on Komadugu Yobe will not fill and no irrigation will be possible. It also appears that the headwater dams will not be helpful during drought periods if their only use is to prevent overflows at Lai and Bongor during the flood season. During a drought such overflows would probably not occur. Therefore, no water would be saved by the dams. The operating policy for the dams during a drought would probably be to let all water pass and flow to Lake Chad to prevent as much as possible the lake level from falling. Irrigation water probably should not be withdrawn.

There is no mention in the LCBC reports of the feasibility of major channellization of the Logone and Chari to prevent losses from infiltration and spills. A typical spill is loss of water by the Logone River to the Benue

River via the Mayo Kebi (Figure 5). There does not appear to be a large potential for upstream storage. Therefore, most of the water would have to be stored in the lake. Levees would have to be built around the lake. If the approximately 5.3 x $10^9 m^3$ spilled, and lost, between Lai and Miltou and N'Djamena were available, the agricultural potential of the area would be greatly increased. Over a year, this would provide irrigation for 140,000 additional hectares. The dams in this case would not be needed. In drought conditions, the total amount of irrigation possible would be about 70,000 ha.

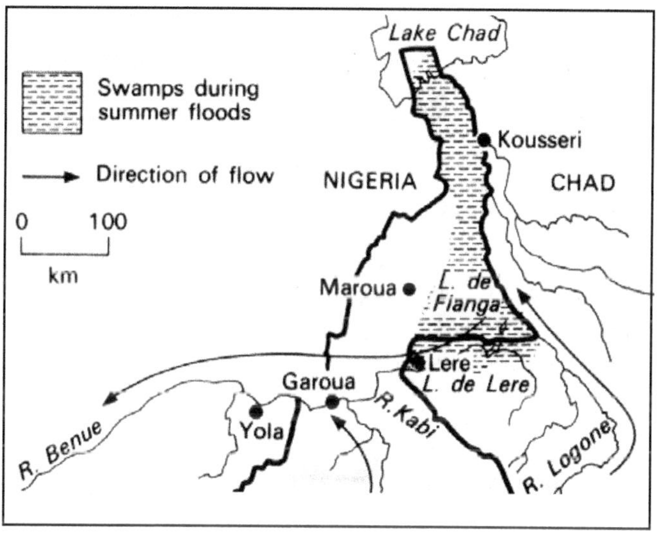

Figure 5: Map illustrating the capture of the headwaters of the River Logone by the Mayo Kebi.

Presently, irrigation in the basin is mainly in the Nigerian portion and Yagoua in Cameroon. There is little irrigation elsewhere. If it is assumed that the maximum amount of permanent lowering of the lake level is one metre, since over 95% of water entering the lake enters from the Chad side, that is, the Logone – Chari system, the volume given could either be entirely allocated to the riparian areas of Chad south of the lake or else distributed between Chad and Niger (assuming Niger takes from the lake). With no regulation and channelization, in an average year the volume that could be withdrawn from the lake is 7.8 x $10^9 m^3$. While channelization of the Logone

– Chari system to reduce overflow losses has not been suggested in the literature, it may be feasible. If done, this will provide an additional 5.3 x 10^9m^3 of water. The total available would be 13.1 x 10^9m^3.

Under drought conditions the lake would split apart as in 1972. Therefore with drought no large water withdrawals are possible from the lake or the Logone – Chari system under natural conditions. With channelization, a volume of approximate 2.6 x 10^9m^3 would be available under drought conditions, most of it between July and November (the rainy season).

Management Strategy under Increasing Human Water Needs and Climate Change

A serious potential threat to the existence of Lake Chad is the anticipated rise in human water needs, mainly due to population growth and increasing development. Pressures caused by water extraction and climate change will almost certainly interact, exacerbating the effects of climate change on the lake ecosystem and possibly increasing human needs further due to increased temperatures and evaporation. Proposed adaptation strategies for human water resource management include:

- "Demand-side" adaptations, such as price incentives for conserving water, enforceable water efficiency standards, and increased irrigation efficiency.
- "Supply-side" adaptations include building more dams (Arnell *et al.*, 2001) and using Integrated Water Resource Management (IWRM) to adapt to increasing water demands. In this process, all stakeholders are included in potential considerations of supply-and demand-side actions before a decision is made, and the situation is continuously monitored and re-evaluated. Unfortunately, IWRM does not consider maintaining aquatic ecosystem function as one of the goals of water management, environmental damage is only included as a potential negative side effect of some actions (Arnell *et al.*, 2001)

Richter *et al* (2003) have therefore proposed ecologically sustainable water management, which strives to protect the ecological integrity of freshwater ecosystems while meeting current and future water needs of man. This strategy has been applied primarily in cases where one or more large dams already exist on a river and there are conflicts between increasing human extraction and ecosystem flow needs, a problem that is likely to become more common in the future due to climate change. Methods of addressing water conflicts commonly include:

- Altering patterns of surface and groundwater extraction;
- Increasing efforts to improve water efficiency;
- Changing temporal patterns of water release from dams. In this case, all stakeholders are involved in the process of estimating ecosystem flow requirements (seasonal base flow, high and low flows and rates of rise and fall), determining current and future human water needs, identifying incompatibilities (seasonal or regional needs), and collaboratively searching for solutions.

Transboundary Rivers and lakes such as lake Chad can make the implementation of the above strategies difficult. This has been the case of the LCBC. With Integrated River Basin Management (IRBM), the needs and expectations of all "water stakeholders" (local community, civil authority, water and fishery resource managers, scientists, conservationists, and representatives of the private sector) from all countries are assessed jointly, a basin-wide authority created, monitoring methods developed, and an adaptive management plan initiated (WWF, 2002). Decisions can be made locally, but must be in accordance with basin-wide strategy. Key aspects in promoting the success of IRBM are:

- Actively promoting public involvement through appropriately-scaled local discussions suited to the target audience.
- Ensuring sustainable funding for the basin-wide authority often through water taxes, reductions in water subsidies, or international funding programmes.

Because IBRM employs adaptive management strategies, it is able to continuously respond to the effects of climate change. In addition, because management is basin-wide, IBRM provides the ability to protect climate change refugia and areas that are particularly vulnerable to climate change, as well as to minimize damaging land use practices and other human stresses that are known to interact with and exacerbate the effects of climate change.

There is an urgent need to protect this freshwater ecoregion in order to preserve the ecological and evolutionary driving forces of biodiversity. It involves protecting relatively large units of water and surrounding land that contain distinct assemblages of natural communities sharing many of the same conditions (Abell *et al.*, 2002). This is important because the present day fish distribution in most of the rivers in Africa has been affected by the pre-historic fish distribution. For example, it is noted that Lake Mega – Chad was connected to the Nile, Niger and Congo river systems about 15,000 years ago (Beadle, 1974). Out of the 179 lake Chad fish species, 106 species are common to the Nile, Niger and Congo systems, 85 species are common to the Chad and Nile systems and 47 species are common to both Lake Chad and Congo systems (Ssentongo, 1979). The difference between Ecoregion Conservation (ERC) and IRBM is that ERC focuses mainly on preserving biodiversity. IRBM emphasizes balancing environmental, economic, and social needs. ERC is well suited in buffering systems against the effects of climate change. It addresses the conservation needs of a wide-range of species in a large area of habitat, and has the ability to address threats that operate across an entire ecoregion with a single, coherent approach (Abell *et al.*, 2002).

Conclusion

In relation to the climate change, freshwater ERC may provide the most comprehensive and holistic approach to providing adaptation options and minimizing anthropogenic stresses. However, IRBM offers a more balanced approach that considers socio economic factors in adaptive management. It may more likely provide long term protection as human water demands increase and human populations experience more direct effects of climate change and extreme climate events. The current management strategy based on IRBM, which is biased towards physical resource conservation, therefore, does not guarantee the biological integrity of the lake basin resources. The LCBC in the management of the Lake Chad Basin has not been guided by ecological principles aimed at increasing resistance and resilience to climate change, thus has failed to combine the IRBM approach and ERC approach in order to derive the benefits of the two approaches.

References

Abell, R.; Thieme, M.; Dinerstein, E.; and Olson, D. (2002) A source book for conducting biological experiments and developing biodiversity visions for Ecoregion Conservation. *Freshwater Ecoregions, Vol. 2,* WWF, Washington D. C. 201p.

Arnell, N.; and Lui, C. (2001) Hydrology and water resources. In: J. Mclarthly; N. Canziana; D. Leary; Dokken and K. White (eds). *Climate change 2001: Impacts, adaptation and vulnerability.* Intergovernmental Pannel on Climate Change, Cambridge University Press, p. 191 – 233.

Ayonghe, S. (2001) A quantitative evaluation of global warming and precipitation in Cameroon from 1930 to 1995 and projections to 2060: effects on the environment and water resources. In: E. B. Eze and C. M. Lambi (eds). *Readings in Geography*, Unique Printers, Bamenda, p. 142 – 155.

Beadle, L. C. (1974) The inland waters of tropical Africa. Longman, London. 365p.

Bootsma, H. A; and Hecky, R. F. (1993) Conservation of the African Great Lakes: Limnological perspective. *Conservation Biology, Vol. 7,* p. 644 – 656.

Carter, R. and Alkaki, A. (1996) Shallow groundwater in the northeastern arid zone of Nigeria. *Quarterly Journal of Engineering Geology, Vol. 29*, p. 341 – 355.

Diamond, A. W. and Hamilton, A. C. (1980) The distribution of forest passerine birds and Quaternary Climate Change in tropical Africa. *Journal of Zoology, vol. 191*, p. 379 – 402, London.

Dresch, J. (1973) Drought over Africa. The United Nations Economic, social and Cultural Organization, *Courier,* New York.

ECA (no date) Chapter 2: The agricultural environment. Economic Council for Africa, Addis Ababa.

ECA (no date) The hydro-atmospheric environment. Economic Council for Africa, Addis Ababa.

Grove, A.T. (1970) Africa south of the Sahara. Oxford University Press, Oxford, 280p.

Hamilton, A.C. (1976) The significance of patterns of distribution shown by forest plants and animals in tropical Africa for the reconstruction of Upper-Pleistocene palaeo environments: a review. *Paleoecology of Africa, vol. 9,* p. 63 – 97.

Hamilton, A.C. (1982) Environmental history of East Africa: a study of the Quaternary. Academic Press, London.

Kadomura, H. (1986) Geomophology and environmental changes in Africa: Case studies in Cameroon and Kenya. Laboratory of Fundamental Research, University of Hokkaido, Sapporo, 299p.

LCBC (1972) Survey of the water resources of the Chad Basin for development purposes: Surface water resources in the Lake Chad Basin. UNDP / FAO, Rome.

Ledger, M. A. (1964) Some hydrological characteristics of West African rivers. Transactions of the Institute of British Geographers.

Major, C.D.; Kirshen, P.H. and Lengyel, Z. (1974) A framework for evaluating long-term strategies for the development of the Sahel Region: Annex 8 – An approach to water resource planning. MIT, Massachusetts, 282p.

Morin, S. (1982) Notes d'hydrographie et hydrologie Camerounaises. *Travaux et Document de Geographie*, University of Yaounde, 64p.

Ngwa, J.A. (1979) A new geography of Cameroon. Longman Group Limited, London, 151p.

Richter, B.D.; Mathews, R.; Harrison, D.; and Wigington, R. (2003) Ecologically sustainable water management: managing river flows for ecological integrity. *Ecological Applications, Vol. 13*, p. 206 – 224.

Schindler, D.W. (2001) The cumulative effects of climate change and other human stresses on the Canadian freshwaters in the new millennium. *Canadian Journal of Fisheries and Aquatic Sciences, Vol. 58, p. 18 – 29.*

Ssentongo, G. (1979) Freshwater fisheries in Africa: *Sixth FAO / SIDA Workshop on Aquatic Pollution in Relation to the Protection of Living Resources: 12 June – 22 July 1978,* Nairobi and Mombassa.

Tyedmers, P. and Ward, B. (2001) A review of the possible impacts of climate change on BC's freshwater fish resources and possible management responses. *Fisheries Centre Research Report, Vol. 9.* p.1 – 12.

WWF (2002) Managing water wisely: promoting sustainable development through integrated river basin management. Available on-line at http//wwwpanda.org/download/freshwater/managing water wisely eng 2. pdf.

Zinyowera, M.; Jallow, B.; Maya, S.; Okott-Ogendo, H.; and others (1998) The regional impacts of climate change: an assessment of vulnerability. In: R. Watson; M. Zinyowera; R. Moss; D. Dokken (eds). *A special Report of IPCC Working Group II,* Cambridge University Press, p. 30 – 84.

Consulted but not cited archival material of the Lake Chad Basin Commission (LCBC):

LCBC (1969) Study of water resources in the Lake Chad Basin, 1966 – 1970. UNDP/United Nations Economic, Social and Cultural Organization.

LCBC (1973) Survey of the water resources of the Chad Basin for development purposes: Vol. 1, Hydrogeological study. UNDP/United Nations Economic, Social and Cultural Organization.

Chapter Thirteen

Climate Change and Sustainability of Traditional Irrigation Systems

Summary

Climate change has a major impact on biodiversity, and therefore also on ecosystem services such as agricultural production and water availability. At the same time the management of biodiversity through tree and forest protection and regeneration, and agroforestry systems can play a fundamental role in mitigating climate change. The chapter uses a combination of primary and secondary data to make observations on the sustainability of water resource development and use, and the implications of climate on water availability and provision of food in dry lands of sub-Saharan Africa. It analyses the problems of agricultural development based on small-scale irrigation schemes using the ecological, economic, social and political components of sustainable development and concludes that with increasing precipitation variability in the form of droughts and floods, traditional irrigation technologies are already foundering and cannot therefore support intensive and extensive cultivation on a year round basis. The paper therefore recommends that adaptations to climate change that can promote longer-term sustainability are necessary. These include the development of small-scale irrigation schemes based on traditional and modern methods of water extraction and use, the development of rain water harvesting techniques, the protection of indigenous trees and forests, the enhancement of genetic crop biodiversity, the development of agronomic soil and water conservation practices and the development of appropriate agroforestry systems. These solutions apart from ensuring sustainable food and water provision can play an important role in mitigating climate change as well as supporting traditional societies in Sub-Saharan Africa in adapting to it.

Key Words: Climate change, Sudan-Sahel biome, agriculture and water availability, ecosystem services, observations, problems, solutions.

Introduction

Out of the 24 ecosystem services assessed by the Millennium Ecosystem Assessment, 15 are in decline. These include the provision of fresh water, marine fishery production, the number and quality of places of spiritual and religious value, the ability of the atmosphere to cleanse itself of pollutants, natural hazard regulation, pollution and the capacity of agricultural ecosystems to provide pest control. (The International Journal for Rural Development, 2008). Current concerns about global warming and desertification and their impacts on environmental resources are concerns in the discourse of sustainable resource development and use. Water is an important resource necessary for the development of agriculture and life. Cunningham and Saigo (2001) record some rivers in the world that sometimes dry up completely because of human diversions and excess consumption. These include the Colorado in the U. S. A, the Amu Dar'ya and Syr Dar'ya in Kazakhstan, the Ganges in India, the Yellow River in China and the Nile in Africa. The drying up of these once mighty rivers is only a symptom of a larger dilemma. In many countries around the world water shortages are increasingly common. Signhonnou (1996) and Molinier *et al* (2000) concluded that the management of the water resources of the Chad Basin of Cameroon is below expectation. The United Nations warns that water supplies are likely to become one of the most pressing resource issues of the twenty-first century (Cunningham and Saigo, 2001). Sustainable water resource development and use therefore is a global concern. Overall social development is only sustainable if no deterioration occurs in the medium and long-term in the ecological social, political and economic sectors. The content of the concept of sustainable development varies according to culture, time and perspective, and can be understood only by adopting an integrated approach (Kohler, 1996; Chifos, 2006). It represents a major challenge to geographers, because geography in particular is the science which can sensibly combine the biological, physical, social and economic aspects. With this in mind, the chapter seeks to make observations on the sustainability of water resource development and use, and the implications of climate change on water and food provision in the dry lands of sub-Saharan Africa

Study Area

The Chad Basin of Cameroon is a semi-arid region. It is located within latitudes 10^0N and 13^0N and longitudes 10^0E and $12^0$50'E. It has a land surface area of 323,310 km^2 with an average population density of 58 inhabitants / km^2. The area is semi-arid with 7 or more months of drought, a mean annual rainfall of about 900 mm in the South and 400mm in the Lake Chad shore. Mean annual temperature is 28^0C with a mean annual range of 7.7^0C (Moby, 1979). The Logone River and its tributaries originate in areas of high rainfall (Figure 1). Upon reaching the region's plains their gradients are sharply reduced and sediments are deposited, creating the swamps of the Logone River flood plain and the Chad Basin.

Three broad morphological units exist, namely, the alluvial plains, Diamare plains and the Mandara Mountains. The relief is largely below 500 m above sea level and is developed on the Quaternary deposits of the Chad Basin colonized by Steppe savannah. It is a mosaic of wooded savannas, thorn bushes, farmlands and natural pastures. These have been intensively degraded by over-grazing, slash-and-burn cultivation and deforestation resulting in runoff, gullying and alluvial processes.

Lake Chad is shared by Cameroon, Chad, and Nigeria. The surface area of the lake has decreased from 24.00km^2 to 12.00km^2 and the average depth hardly exceeds 4m. The volume of water in the lake is shrinking due to alluviation, evaporation, infiltration and losses due to natural overflow at Bahre-el-Ghazal, which drains northwards into the Bodele depression (LCBC, 1972). This raises concerns on the sustainability of water resources in the Chad Basin. Global warming and desertification also call for an assessment of the sustainability of the water resources, their rational development and usage. Farmers practise rainfed agriculture supported in some cases by traditional irrigation.

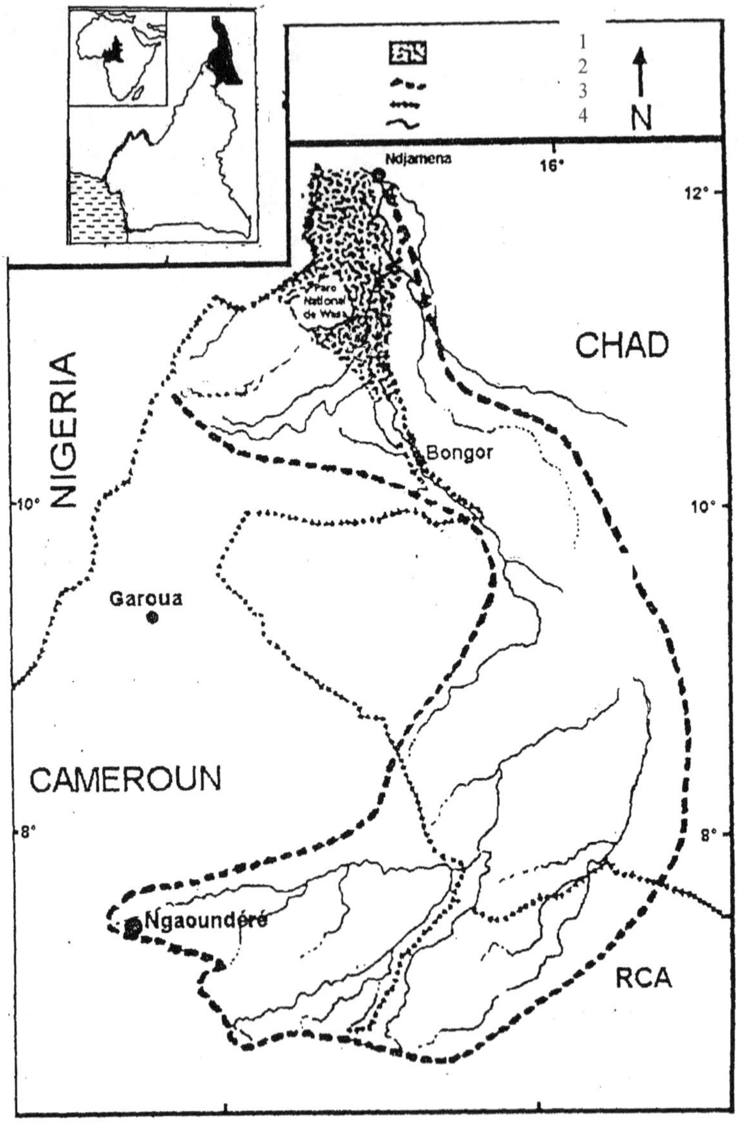

Figure 1: *Location of the study area: The Chad Basin of Cameroon: 1 = seasonally flooded zone, 2 = drainage basin limit, 3 = frontier, 4 = rivers.*

The discourse of sustainable development is concerned with identifying models of development that facilitate the equalization of benefits across time and space. This necessitates the rethinking of the relationship among natural, social, cultural, political, economic and built systems as they act and are acted upon along a pathway of change (Chifos, 2006). The discourse opens an arena to re-examine our conceptions of progress and to re-define trajectories that will enable that progress for more people both today and long into the future (Meadowcroft, 1999).

Study Methods

The study collates data obtained from primary and secondary sources to assess current water development and management strategies as they relate to questions of sustainability. Primary sources include field observations and irrigation – related land use activities in three villages using aerial photographs from the National Geographic Institute (IGN Mission 67 – 68 CAM 01/200 No. 189 and IGN Mission 78 CAM 131/200 No. 54), that is, Doreisso, Doumaraye and Vounaloum. Data on climatic parameters investigated were obtained from the Provincial Meteorological station in Maroua. These were complemented by secondary data. Hydrological characteristics were obtained from the archival material of SEMRY (Yagoau Rice Development Authority) and other documented sources. Field observations in the three villages yielded data on water use techniques and water resource potentials. The data so obtained were used to assess issues of water development and management related to the ecological, economic, social and political components of sustainable development, as they relate to climate change.

Problems of Water Resource Availability and Use

The region is characterized by a Sudano-Sahelian climate with 7 dry months and a 5 months wet season. July and August rainfall accounts for 66% of the total annual rainfall. Inter-annual rainfall variability is also typical. (Figures 2 and 3). Inter-annual average rainfall before 1970 was 700mm. After 1970 it dropped to 500mm following a general fall in the rainfall regime in the whole of tropical Africa (Olivry, 1986; Mahe and Olivry, 1991; Olivry, 1993 and Olivry *et al* 1994).

Table 1: Frequency of Variability of Annual Rainfall (1970 – 1988) and mean annual total number of rainy days (1940 – 2060)

Probability of rainfall greater than the threshold indicated	Threshold rainfall (mm)			
	Kaele	Maroua	Garoua	Touboro
0.8	622	631	820	987
0.5	655	675	883	1054
0.2	781	780	1014	1247
Average for the period	666	752	951	1180

Mean total number of rainy days computed from trend lines						
Stations	1940	1960	1980	1995	2030	2060
Maroua	72.8	71.8	70.8	70.1	68.4	66.9
Garoua	82.1	79.7	77.2	75.4	71.1	67.4

Table 2: Evaporation Rate in some Stations

Station	Evaporation (mm/month)												
	J	F	M	A	M	J	J	A	S	O	N	D	**Total**
Poli	236	279	323	225	127	66	55	56	51	81	149	202	**1850**
Maroua	228	296	365	304	203	199	81	60	83	143	229	213	**2396**
Kousseri	345	389	506	451	362	265	149	74	96	206	353	341	**3519**

Source: Maroua: Provincial Meteorological Station

Table 1 presents the frequency of rainfall variability for the period 1970 to 1988 and the mean number of rainy days computed from trend lines (Ayonghe, 2001) for the period 1940 to 2060. According to Ayonghe (2001) high rainfall was evident from 1951 to 1967, 1977 to 1980, and 1989 to 1995, while low rainfall was observed from 1930 to 1950, 1968 to 1976 and 1981 to 1988. In general the mean total number of rainy days for the period 1940 to 2060 is on the decline. The rainy season lasts three to six months depending on the location and it is characterized by inter-annual variability of rainfall. Figure 3 presents the frequency of the total annual rainfall with a probability of 8 out of 10 years for the period 1952 to 1969 and 1970 to 1989. It again depicts inter-annual variability.

The consequences of this climatic situation are a biological rhythm with two phases:

- An active vegetative phase of very intense growth of about 5 months. The reconstitution of ground water reserves is essentially achieved from July to August when rainfall exceeds evapotranspiration. This reserve is rapidly depleted as soon as the dry season sets in (November to May). Evaporation rates are high (Table 2)
- The wet season is followed by a dry and hot season during which ground water reserves are exhausted by high rates of evaporation. The physiological activities of plants are very restricted and much of the vegetation dries out.

Streams and rivers are characterized by irregular flow rates. The plains consist of numerous small elongated, sub-water sheds that are parallel to the broad alluvial plain of the Logone River. These discharge runoff water and deposit sediments in a network of outwash splays. The typical hydrological features are:

- An intermittent flood flow during the wet season (June to October) and total drying up of streams and rivers during the dry season. (November to May).
- The disappearance of channel characteristics in downstream areas where the gradient drops (December to May); and
- Deposition of large volumes of sediment in the vast flood plain of the Logone River. The river regime is determined by the

rainfall. Fulton *et al* (1974) observed that the normal discharge of Logone River in 1974 was 65m^3/s while the peak discharge was 92 m^3/s. Molinier *et al* (2000) observed that the decrease in rainfall between 1970 and 1984 was reflected in the drying up of the river. In 1984 the average discharge was 551 m^3/s, the flood discharge was 2740 m^3/s while the low stage was 50 m^3/s. The inter-annual variability of the discharge is a reflection of the rainfall characteristics.

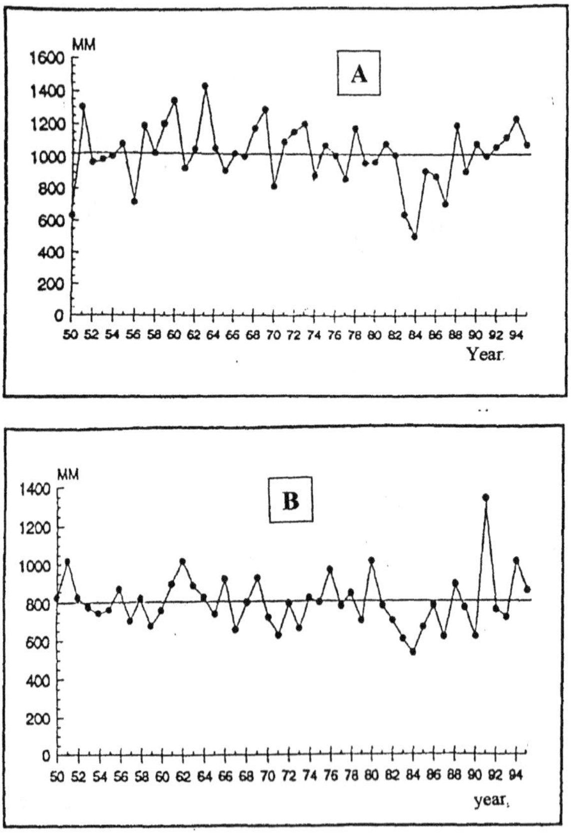

Figure 2: *Evolution of annual rainfall in some stations: A = Garoua from 1950 to 1995 and B = Maroua from 1950 to 1995 (after Donfack, Boukar and M'Biandoun, 1996)*

The lower section of the Logone River consists of a broad, level plain. A slight increase in the rate of channel flow causes over flooding of the stream.

This allows water to spread over the broad alluvial plain behind natural levees. Widespread flooding results and a large volume of water is exposed to evaporation. Annual average evaporation rates in some stations are: Poli 1815mm/year, Maroua 2396mm/year and Kousseri 3519mm/year.

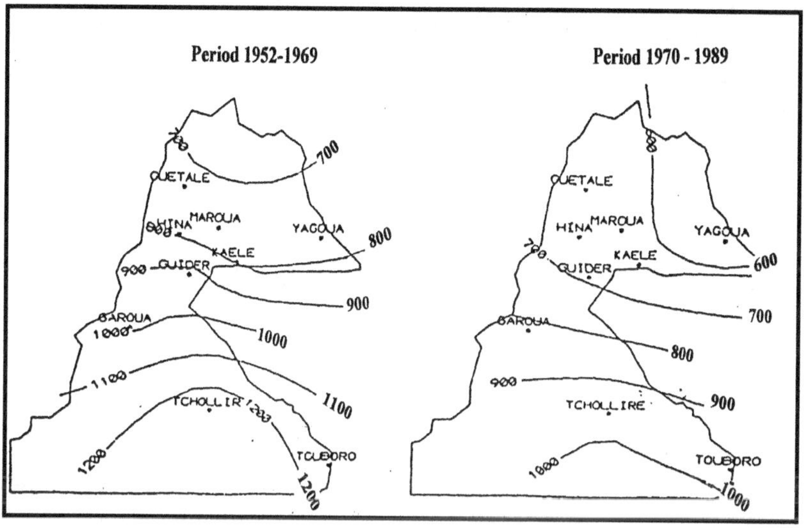

Figure 3: Frequency of total annual rainfall (mm): 8/10 years (After Mbiandoun, 1990).

The flooding of the alluvial plain commonly gets to a height of 80 – 120 for a period of 3 months in broad reaches and continuously in enclosed depressions (Beauvilain, 1985; Roupsard, 1985). This provides a reliable supply of water for irrigated rice in March and April, that is, during the low stage. A number of interrelated factors are important in considering drainage measures to increase the productive capacity of the area. Benefits from such measures include:

- Rice could be effectively grown where flooding is not excessive or where it can be controlled;
- Management of receding surface water will allow cultivation of dry season sorghum in a non-rainfall period;

In the early 1970s the seasonally flooded areas of the Logone flood plain experienced a 60% reduction in their land surface area. This was attributed to droughts (Fulton, *et al.* 1974). This adversely affected agriculture and fishing.

There was crop failure due to the lack of rainfall. Rivers also dried up. The government of Cameroon was obliged to modify the traditional use of water resources in the region through the creation of a water management authority known as the Yagoua Rice Development Authority. It was charged with the development of irrigated rice in the area based on the management of flood water and the construction of a reservoir at Maga in 1979. The reservoir has a capacity of 600 million cubic meters of water and covers a surface area of 39.000hectares.

Studies by Naah (1990) reveal that despite the construction of the reservoir the water requirements for agriculture have not been met for about 25years due to the continuous drop in the volume of rainfall. The recharge of groundwater resources results from the Logone River, flood waters and the seasonal streams (Naah, 1990). It is therefore logical to conclude that the absence of considerable flood waters contribute to a reduction in phreatic water necessary for dry season irrigation. Farmers depend on traditional forms of irrigation, namely:

- Using the rainy season floods to grow crops as the flood waters recede;
- Water may be obtained from streams, rivers and natural ponds and stored in reservoirs for eventual distribution via canals;
- Water may also be drawn from natural ponds or rivers by use of the "shadouf" system of irrigation;
- Alternatively, water can be drawn directly by hand from wells or natural ponds and poured into canals. A large number of diversion systems and canals have been constructed;
- Under the water management authority water is pumped into canals for distribution through a network of canals and dykes.

The irrigation technique used is mainly flood irrigation. The techniques used by farmers have remained unchanged for centuries. It is therefore a system embedded in the indigenous knowledge and culture of the local people.

Problem of Climate Change and the Sustainability of Water

The Millennium Ecosystem Assessment (http:/www.millenniumassessment.org) was the first global assessment of changing "ecosystem services" as a consequence of global environmental degradation. Ecosystem services describe functions of ecosystems that are valuable for man and human development. In this case, it is the provision of fresh water. One of the greatest challenges for rural development and poverty alleviation is to tackle increasing variability in precipitation which is characterized by floods and droughts, combined with an overall loss of biodiversity not only due to climate change, but also due to over exploitation resulting in soil erosion. This section presents the main observations on the complex relationships between climate change, water and food provision.

The Lake Chad Basin Commission (1972) reports that by 2020 an area of 260.000hectares can be irrigated using water from Lake Chad sources. It is estimated that this will require 3.9 x 10^9m^3 of water, or 0.5 litres /sec /ha. If we assume year round cropping at 1 litre/sec/ha. This will lower the equilibrium level of the lake by 76cm. The potential from the rest of the basin totals 18.000 ha, much of which is seasonal. In a drought, it is expected that most of the seasonal agriculture would be eliminated. Hydrologists have advised against the large scale use of artesian ground water to avoid depletion of reserves that are a part of the delicate balance sustaining Lake Chad. Any plan for the use and management of deepwater sources should take into account the extravagant use of groundwater reserves in the Chad lowlands in neighbouring Nigeria. Many wells are only a few kilometres outside the national frontier of Cameroon.

Grainger (1986) estimated that every additional person in dry lands requires one more hectare of land to supply him or her with the 250kg of grain needed for subsistence per year. With a current population growth rate of 3.8 percent the demands made on the water supply clearly reveal the disadvantages of traditional irrigation technology. With the construction of highways, the Trans-Cameroon railway and the re-opening of the Garoua river port, economic links have been increasingly forged beyond the boundaries of the region. This has promoted the cultivation of cash crops for which an irregular or only seasonal supply of water is neither adequate nor suitable. The problem is also compounded by limited available potential agricultural land. Overgrazing, deforestation of thorn bushes, bush fires, slash-and-burn shifting cultivation and land use pressure. These together with droughts and floods exacerbate the desertification processes and soil erosion.

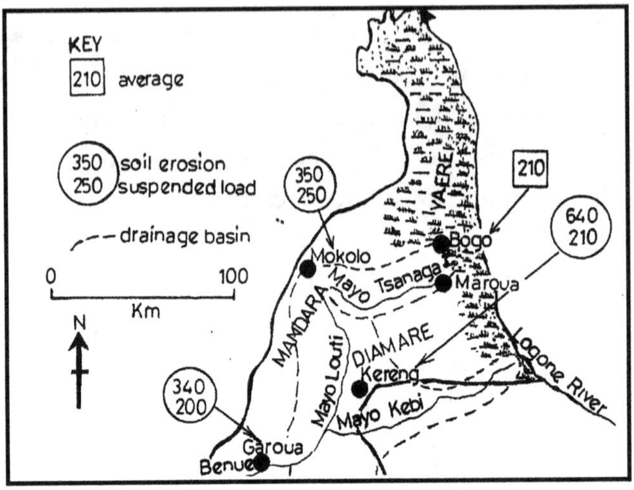

Figure 4: *Annual soil toss (tons/ km² / year) in the Chad Basin of Cameroon (Olivry 1978)*

Figure 4 presents the mean annual erosion rates. The Mayo Tsanaga River at Bogo Station has a suspended solid transport load in the range of 325,000 to 335,000 tons/km^2/year. Mayo Louti River has a suspended sediment load of 250 tons/km^2/ year and Mayo Kebi River about 210 tons/km^2/year. These are all seasonal rivers. The only perennial surface water is the Logone River with a suspended sediment load of about 630 tons/km^2/year. The soil loss in the river basins increases with the extent of cultivation, grazing and annual bush fires. Mean soil loss in these watersheds ranges between 310 to 630 tons/km^2/year. Climate change with associated droughts and floods will cause more soil erosion.

Table 4 presents the discharge of water in various watersheds during the dry season and the wet season. But for the Logone River the others are seasonal. The channel flow of the Logone River forms the basis for small-scale irrigation schemes. The low discharge stage, which occurs from March to April, provides a reliable supply of dry season water for downstream irrigation. Flooding of the alluvial plain is commonly to a height of 120 mm for a period of 3 to 6 months in broad reaches and continuously in enclosed depressions. This provides water for a significant craft fishing industry and ensures a necessary source of food. Climate change with associated floods and droughts will reduce the volume of water available for ground water recharge (Figure 5):

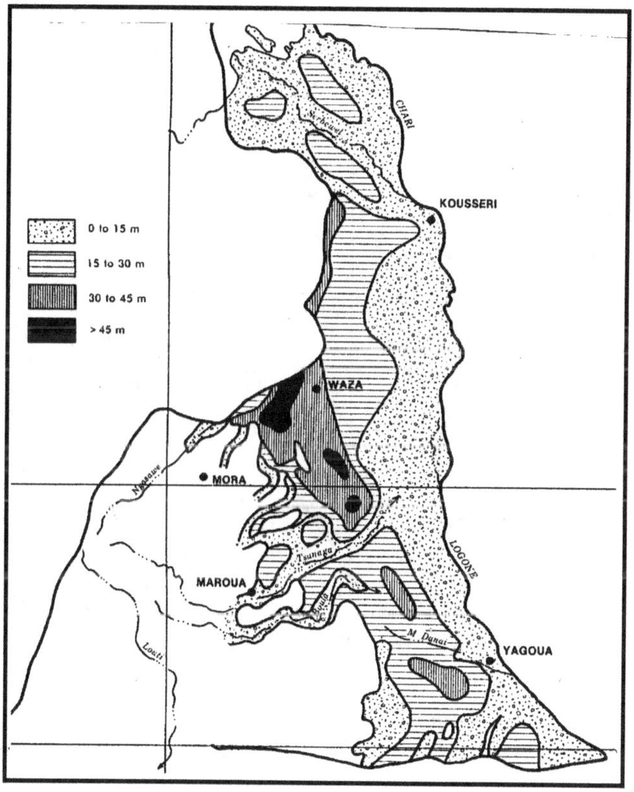

Figure 5: *Phreatic water table in the Chad Basin of Cameroon (Fulton et al, 1974)*

Table 4: *Water discharge in watersheds (cubic millimetres per second after Fulton et al, 1974)*

River	Station	Lower water stage	Flood stage
Benue	Garoua	1	2,375
Logone River system:	Wakwa	7	164
Vina	Maroua	Underground	228
	Bogo	Underground	210
Kalliao	Yagoua	50	905
	Kousseri	45	920
Tsanaga			
Logone			
Logone			

- Ground water resources are non existent or impractical to develop in areas of high relief (Mount Mandara and Diamare plain).
- The potential for the development of ground water is moderate at an elevation of less than 250m above sea level.
- Regions adjacent to the Logone River flood plain have increasing potential northwards with distance to the present shores of Lake Chad (Yagoua-Kalifou, Doukoula, Maroua-Bogo)
- Spring sources are occasional and meagre in dissected areas of the Mandara Mountains. Some of these have been developed for seasonal use.

The management of soil and water resources in dry lands must therefore target the maintenance of both soil fertility and structure, and the replacement of both surface and ground water resources. The relationship between vegetation, rainfall, surface water and ground water recharge constitutes a bundle of ecosystem services.

Traditional irrigation in the basin depends on the occurrence of floods. In drought years farmers depend on groundwater. Access to water for irrigation becomes difficult due to the high cost of drilling boreholes and procurement of diesel-powered pumps. Water delivered by pumps is applied to crops using flood irrigation techniques. Kohler (1996) notes that this modern and expensive method of pumping of water, combined with an inefficient distribution method, can exist only for as long as sufficient quantities of the water resource are available. More efficient systems of water distribution such as drip irrigation are rarely used, because with apparently plentiful reserves of groundwater, there is no urgent need to invest in such technology. Pump irrigation also means that farmers are no longer dependent on the periodic supply of floodwater, because groundwater is available all year round. Thus the merit of adopting pump irrigation as an innovation is twofold: to intensify agriculture and to eliminate the risk of unpredictable floods. The regular and reliable supply of ground water in peri-urban areas and for cash cropping in rural areas has resulted in dynamic, economic development. But, if this development is to be seen as an improvement for future generations, capable of providing inter-generational equity, it should not accelerate global warming and desertification and deplete the natural resource base in the short-term without providing an adequate substitute for these resources.

An analysis of future prospects must focus either on short-term economic development or on long-term sustainability in all sectors:

- Ecological sector: Traditional irrigation systems in the area use resources in an ecologically sustainable manner, because the rate of water consumption cannot be influenced. No over-use is possible but some changes may occur in the water budget of the basin. On the other hand, irrigation systems based on the pumping of ground water have a high degree of water over-use, salinization of the soil and of the remaining ground water. Lake Chad; an inland drainage lake is already shrinking in size. The lake Chad Basin Commission (LCBC) charged with the management of the natural resources in the area is trying to strike a balance between water withdrawal and regeneration. This requires limiting the amounts of water withdrawn; a measure which is being frustrated by inefficient irrigation systems.
- Social sector: Traditional irrigation systems depend on common access to surface water resources whose impact on sustainability is not possible to assess. Despite the risk of a "tragedy of the commons" situation, these systems are very compatible with social structure and indigenous knowledge base or culture. On the other hand, modern irrigation systems based on groundwater extraction are not compatible with the culture of the people. Surface water sources are common access resources while access to groundwater sources depends on land ownership and the ability to procure diesel pumps.

- Political sector: Traditional water use systems correspond to the political structures and institutions of the ancient kingdoms. Ancient kingdoms own and use water resources. Access to these resources depends on the strength and structure of indigenous political institutions and the social and political status of individual members.
- Economic sector: Traditional systems of water use are characterized by inefficient use of capital and labour, and no possibility of development. All the systems have negative and positive aspects. Traditional systems have a high long-term sustainability but are capital and labour intensive. Modern systems require large investment. Modern systems require large capital investments in the construction and operation of pumping equipment. However, the cultivated crops are of high value, hence they are more marketed.

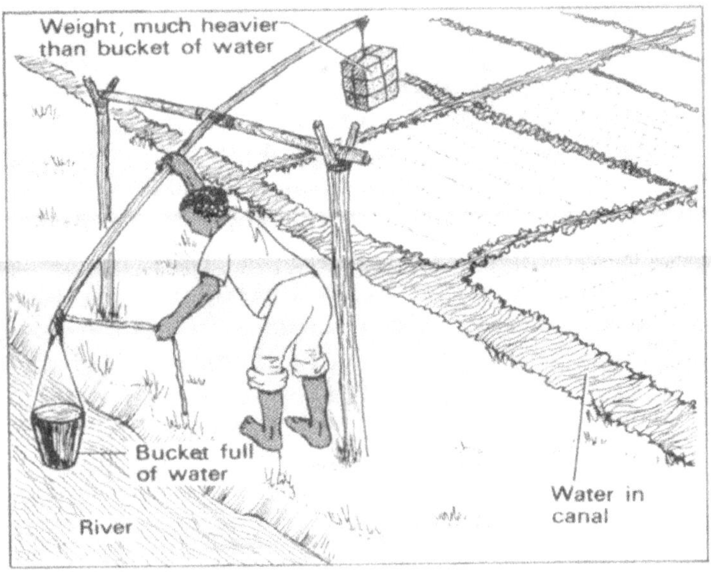

Diagram: *Diagram of the shadouf; used for irrigating cotton and rice fields in the Logone River flood plain (J.A. Ngwa, 1979): Traditional irrigation system.*

Photo: *Modern irrigation in the Chad Basin: Mobile Sprinkler System (Ndenecho): Based on pumps and boreholds.*

Photo: *Flood irrigation or basin irrigation in the Chad Basin (Ndenecho): Based on surface water source.*

Which irrigation technology will therefore guarantee sustainable development? The traditional irrigation methods are already foundering in the face of the actual socio-political realities. These systems cannot satisfy the demands for water delivery capable of supporting intensive and extensive cultivation on a year round basis. The possibility to create added value from agricultural production to help support economic development is limited. Pumped irrigation permits all year round cultivation of cash crops (rice, cotton, groundnuts, onion and beans) which are marketed outside the region. If the yields of such a diversified agricultural economy are high and the irrigation techniques made more efficient, one can talk of sustainable development despite overexploitation of the water resource. Sustainable water use in the social, political and economic sectors could best be achieved by adopting a combination of flood irrigation and pump irrigation. This is possible because the lower section of the Logone River consists of a broad, level plain. With excessive rise of water from headwater sources. Water is diverted by natural spillage across the low saddle into the Benue River by way of the Mayo-Kebi. This causes an annual loss of water from the Logone supply of 40million cubic metres. Investments to harvest this spillage by building reservoirs will compensate for the lack of water during the dry season. Diesel – pumped water will be delivered to the same fields during the dry season in order to extend the agricultural calendar, this will result in advantages of traditional methods (ecological sustainability, low investment and operating costs) being combined with the benefits of diesel – pumped irrigation (all year round reliable supply of water), without excessive over – use of the water resource.

One approach to combine the requirements of ecosystems and agroecosystems for water is rainwater harvesting schemes combined with small-scale irrigation, based on new as well as on traditional methods. In Africa, a large potential exist to increase and safeguard yields in a sustainable way.

Another link between climate change and agriculture lies within the area of soil and water management. Both on the surface and underground, vegetation contains large amounts of carbon. Soil tillage in combination with high temperatures leads to a rapid oxidation and loss of soil carbon and soil moisture (Vohland and Lotse-Campen, 2008). Organic farming and soil conservation practices enhance both the biological diversity of the soil fauna, soil moisture content and reduce carbon losses from the soil. Maintaining soil ecosystem services allows a more sustainable agriculture to face climate change.

The adoption of sustainable water use practices, water harvesting and soil conservation practices must be accompanied by agroforestry and the protection of trees and forests. One of the key roles that vegetation plays with respect to climate change is buffering against increasing temperatures and increasing the probability of precipitation. Agroforestry systems are another possibility to combine biodiversity conservation and climate change adaptation. They are more resilient against climate change than tree-less fields, as the trees reduce wind speed and therefore evaporation, and trap water and nutrients. Agroforestry systems fulfil a bundle of ecosystem services and are therefore important in climate change adaptation.

Climate Change Impacts On Crops

At the level of the crops grown in sub-Saharan Africa, climate change will gradually affect the phenology, that is, the time of the year when crops flower and develop fruits. The increased variability of rainfall might decrease the number of plant generations per year. Increasing temperatures will combine with decreasing or at least more variable supply of water. Another climate change-related impact is the potential "fertilization effect" of the increased atmospheric Co_2, therefore increased supply may lead to increased biomass accumulation or enhanced phenological development – as long as water and other nutrients are not limiting. There is a need to enhance the genetic crop biodiversity. High biodiversity allows for local adaptation processes. Supporting farmers in maintaining their logically adapted native varieties is an important insurance against future increase of temperature variations and the occurrence of new pests and diseases. It is an even better safeguard than genetically modified crops, which often have a reduced genetic variability.

Conclusion

The hydrological resources of the Chad Basin of Cameroon like most of the Sudan-sahel areas of Africa exhibit inter-annual variations and a general tendency towards shrinking groundwater and surface water resources. Water supplies are increasingly becoming a pressing resource issue. The concept of sustainable development is a vision which delivers the general orientation with which operational goals can be defined in specific situations. The soil and water resource management strategies and crop breeding programmes of sub-Saharan Africa must seek to establish the link between climate change and ecosystem functions. More holistic and longer-term orientations are required to assist in negotiating the operational goals. Key areas for adaptation to climate change include:

- The development of small-scale irrigation schemes based on new as well as on traditional methods of water extraction and use.
- The development and promotion of rain water harvesting technologies.
- The protection of indigenous trees and forests.
- The development of appropriate agronomic soil and water conservation practices.
- The development of appropriate agroforestry systems.
- Enhancement of genetic crop biodiversity.
- Breading for drought resistance

Acknowledgements

The author acknowledges the use of material from the article titled "An assessment of the sustainability of water resource development and use in the Lake Chad Basin of Cameroon" published in the Journal of the Cameroon Academy of Sciences vol. 7 (2008) by Dr. E.N. Ndenecho.

References

Ayonghe, S. (2001) A quantitative evaluation of global warning and precipitation in Cameroon from 1930 to 1995 and projections to 2060: Effects on environment and water resources. In: C.M. Lambi and E.B. Eze (eds.) *Readings in Geography*. Unique Printers, Bamenda p. 142-155.

Beauvilain, A. (1985) Remarques Sur la situation pluviométrique actuelle au Nord Cameroun. In ; *Cameroon Geographical Review*, Vol. 5, No. 1 Yaounde University. p. 47 – 62.

Chifos, C. (2006) Culture, environement and livelihood: potential for crafting sustainable communities in chiang Mai. Int. *Journal of Environment and Sustainable Development,* Vol. 5, No. 3.

Cunningham, W. and Saigo, B. (2001) Environmental Science. McGraw Hill, san Francisco

Donfack, P; Boukar, S. and M'Biandoun, M (1996) Les grandes caractéristiques du milieu physique In: Agricultures des savannes du Nord Cameroun. Boukar et al (eds), *Actes de l'atelier d'échange.* CIRAD/CA p. 29 – 40

Fulton, D ; Brower, j ; Boulet, J. *et al* (1974) Resources inventory of North Cameroon – Africa. USAIS-FAC (France) p. 64 -67

Grainger, A. (1986) Desertification. Earthscan Publication, International Institute for Environment and Development, Washington D. C. p. 25 – 26.

Kohler, S. (1996) The demands and limitations of sustainable water use in arid lands: a discussion of "sustainable development" taking the Wadi Markah (Republic of Yemen) as an example. In: *Applied Geography and Development,* Vol. 47. Institute for Cooperation, Tubingen, George Hauser Press, Metzingen, p. 25 – 36.

LCBC (1972) Survey of the water resources of the Chad Basin for development purposes: surface water resources in the Lake Chad Basin, UNDP/FAO Rome.

Mahe, G. and Olivry, J. (1991) changements climatiques et variations des écoulements en Afrique Occidentale et Central, du mensuel a l'international. In: *Proceedings of Vienna Symposium.* IASH No. 201.

Meadowcroft, J. (1999) Planning for sustainable development: what can be learnt from the critics. In: M. Kenny and J. Meadowcroft (eds.) *Planning Sustainability*, London, Routtage. pp 12-34.

Moby, E. (1979) Climate. In: J-F Loung (ed.) *Atlas of the United Republic of Cameroon.* Editions Jeune Afrique. pp.16 -19.

Molinier, M; Sighomnou, D. and Sigha, K. (2000) Perturbation du milieu naturel du Yaere dan le Nord Cameroun: changement climatiques ou action anthropiques? In: F. Kengne and G. Courade (eds) *Societés et environnement au Cameroun.* NGCC, The University of Yaounde I. p 155 - 170.

Naah, E. (1990) Hydrologie du grand Yaere du Nord Cameroun. Unpublished Doctorate Thesis. Faculty of Science, University of Yaounde.

Olivry, J.C. (1978) Suspended solid transport load in Cameroon. In: *ONAREST scientific papers,* vol. No. 1. Yaounde. p. 47 -60.

Olivry, J.C. (1986) Fleuves et rivières du Cameroun MESRES – Yaounde / ORSTOM. *Coll. Monographies hydrologiques,* No. 9. 733p.

Olivry, J.C. (1993) Vers un appauvrissement durable des resource actuelle sur l'écoulement du fleuve sénégal et l'hypersalinité de la Basse-Casamance. In: *Proceedings of the Vancouver Symposium,* August 1987, IAHS, publication No. 168.

Olivry J.C.; Briquet, J. and Mahe, G. (1994) De l'evolution de la puissance des crues des grands cours d'eau intertropicaux d'Afrique depuis décennies. *Dossier de la Revue Geographie Alpine,* No. 12.

Sighomnou, D. (1996) Restauration hydrotechnique de plaine de Logone dans Extreme-Nord du Cameroun. *Prévision des Inondation,* Yaounde.

The International Institute for Rural Development (2008) what means "biodiversity" *The International Journal for Rural Development*, vol. 42, No. 2 p.2

Vohland, K. and Lotze-Campen, H. (2008) Biodiversity and climate change. *The International Journal for Rural Development.* Vol. 42, No. 2 p, 11-13

Chapter Fourteen

Climate Change and Development of Rainfed Agriculture

Summary

Climate change and variability constitute multiple threats to regions dependent on rainfed agriculture. Studies of vulnerability to climate change are needed to raise the awareness, understanding and interest of planners, decision-makers and policy makers in designing adaptation measures. The study uses a combination of primary and secondary data to design a framework for political reactions, public awareness raising and educational measures to be taken to make adaptations to climate change in Cameroon. It posits that despite the vantage position of climate change in international politics and media, the political will to counter climate change in Cameroon is yet to be developed. However, the climate's direct impact on sustainable livelihoods obliges farmers to adopt new cropping practices and strategies to lessen the adverse effects of climate variations on crop yield, farm profit and household income. Finally, the chapter concludes that these indigenous adaptation strategies can be supported and enhanced by appropriate policy actions. Policy directions include adaptations strategies involving crop water requirements, enhancement of indigenous adaptation measures, alleviation of barriers to adaptation, water management to reduce inter temporal and spatial variability of stream flows, careful selection and diversification of adaptable crop varieties and animal breeds, and the promotion of irrigation and rain water harvesting technologies. These hinge on the farmers experience, availability of extension advice specifically related to climate change, enhanced farmer education and good farm-to-market roads to facilitate rural-urban linkages and innovation diffusion. Sustainable adaptation requires institutional and human capacity building, inter-sectoral and ministerial collaboration, developing partnerships with related international institutions and good governance. These must be an intrinsic part of national legislation as well as government and administration.

Key Words: Climate change, Cameroon, political reactions, public awareness, educational measures, agricultural development, policy implications.

Introduction

Osman-Elasha (2007) observed that one of the constraints to understanding the current and future climate variability is lack of sufficiently dependable observable climate data in Africa. The low level expertise in climate science has led to a low adaptive capacity to climate change (Aliyu, 2008). Extreme changes include global warming, drought, desertification, floods, sea level rise, windstorms, rainstorms and thunderstorms among others. Various researchers have studied the causes and effects of extreme weather conditions as they relate to climate change (Awosika *et al.*; 1992; Adelekan, 1998; Nyelong, 2004; Odjugo, 2000; NEST, 2004). On the other hand, there is the lack of information on the economic impact of climate change. The few known studies include Odinga (2004), Buadi and Ahmed (2006), Malun (2006), Reid *et al.* (2007) and Fondo (2001). Other researchers have attempted to quantify the impact of climate change on natural resources in Africa (Velarde *et al.*, 2005; Macgregor *et al.* 2006; Fondo, 2001; Ayonghe, 2001). Ariel *et al.* (2008) researched on the impact of climate change on agriculture in Africa with a focus on impact assessment and adaptation strategies.

Climate change is studied under various facets namely; the science, vulnerability, adaptation and mitigation. Vulnerability refers to the extent to which climate may harm or damage a system. Since this area has received little research attention, decision-makers in Cameroon have little knowledge on how vulnerable Cameroon is to climate change. In-depth studies of vulnerability to climate change that can reveal the sensitivity of a system's sensitivity are needed to raise the awareness, understanding and interest of planners, policy and decision makers. Such an understanding and awareness is necessary in making adaptations, that is, an adjustment in natural and human systems in response to actual or expected climate stimuli or their effects, which moderates harm or exploits beneficial opportunities (IPCC, 2001). Climate change adaptation requires the use of good agricultural, forestry and fishery practices to meet changing and more difficult environmental conditions and introduction of improved risk management measures (FAO, 2007). This chapter uses Cameroon as a case study to design a framework for political reactions, public awareness and educational measures to be taken to make adaptations to climate change.

Problem Setting

Climate change impacts in Cameroon are real especially if one considers the fact that over 70% of the population is dependent on agriculture. Cameroon has a total land area of about 475,440 km^2 with a population of 17 million inhabitants and a growth rate of 2%. Agriculture contributes 44% of the Gross Domestic Product (GDP). Because physical conditions in Cameroon vary, cropping systems and types of crops cultivated vary throughout the country. Three agro-ecological regions can be identified namely; the humid, moist forest zone, the high savannah zone and the sahel savannah zone. Current findings by Ariel *et al.* (2008) in Cameroon indicate that even though farmers may be adapting to the climate variations; the government needs to help the adaptation process by making available the necessary resources and providing irrigation infrastructure, especially in the dry lands of North Cameroon, to counter the debilitating effects of low moisture, peaking daily temperatures and runaway evapotranspiration. To ease the problems of available water deficiency and to enhance crop productivity there is a need to modify cropping patterns and cultivate crops with lower water requirements, and improve irrigation by changing the traditional systems to more efficient systems.

The above measures are necessary to avert the net revenue fall resulting from the decrease in precipitation or an increase in temperature.

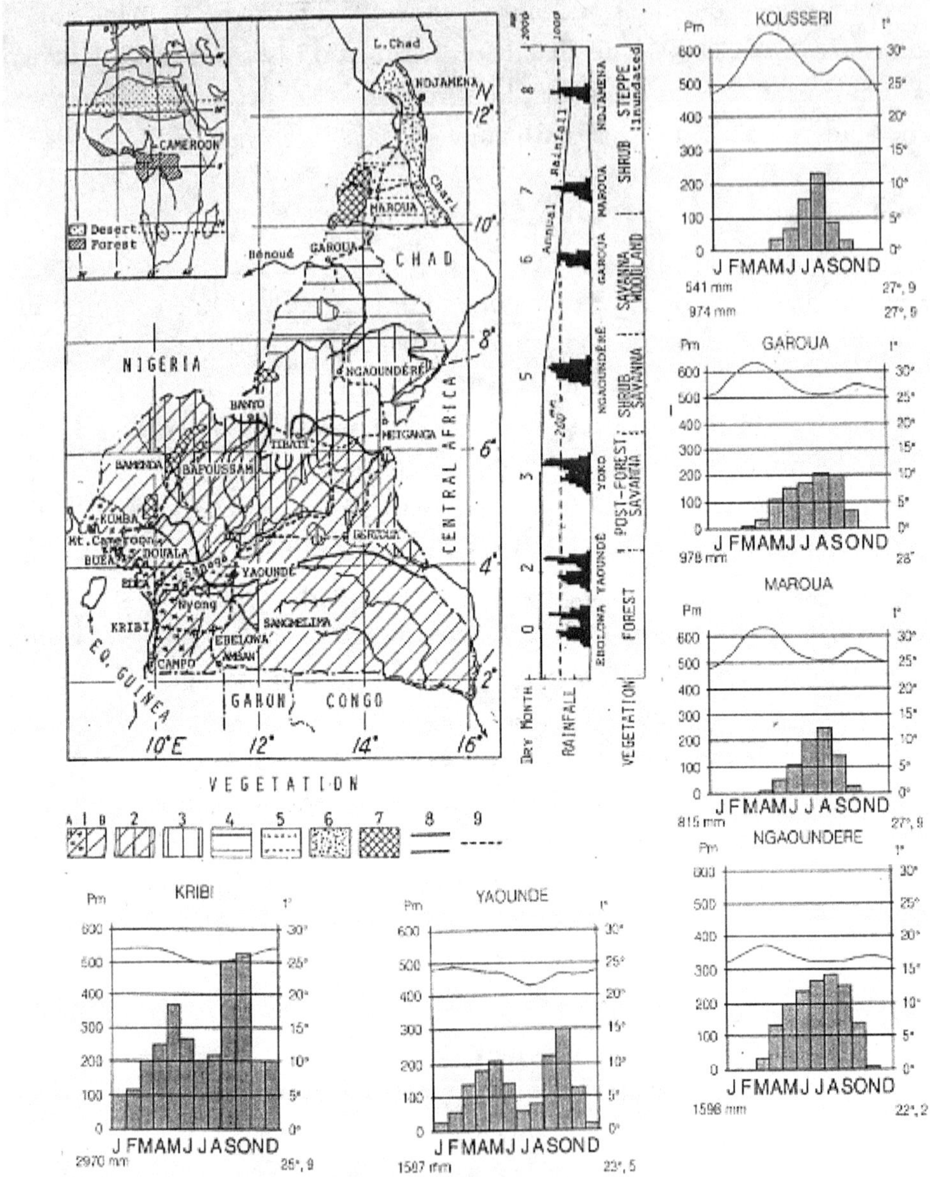

Climate, vegetation and physiography of Cameroon. Vegetation (after LETOUZEY, 1959,1979) Dense humid evergreen forest, including mangrove, 1B: Dense humid semi-deciduous forest, 2: Post-Forest savannah (grass and shrub savannas), 3:Shrub and grass savannas of Adamoua Plateau, 4: Tree savannah woodland and dry woodland (Sudanese zone), 5: Tree and shrub steppes with thorn bushes (Sahelian zone), 6: inundated periodically, 7: Montane and subalpine formation, 8: Zone of ancient and recent destruction of dense humid forest, 9: Main survey routes of 1980/81 season field work.

Figure 1: *Climate, vegetation and hydrography of Cameroon.*

Scientist estimate that an increase in temperature of 2.5^{0}C in Cameroon will cause a net revenue from farming to fall by 0.5 billion dollars. An increase of 5^{0}C will cause a net revenue fall of 19.5 dollars / hectares and a 7% decrease in precipitation will cause a fall in net revenue from crops to 26.8 dollars / hectare. A 14% reduction in rainfall would cause significant losses to the economy. This results from Cameroon's dependence on rainfed agriculture (Table 1 and Table 2).

Table 1: *Marginal impacts of climate net on revenue in Cameroon (dollars/hectare)*

S/N	Variable	All farms
1	Temperature	- 15.40
2	Precipitation	5.65

Source: Ariel et al., 2008

Table 2: *Impacts from uniform climate scenarios in Cameroon*

S/N	Impacts	Increased warming		Decreased precipitation	
		2.5^{0}C	5^{0}C	7%	14%
1.	ΔNet revenue (dollars / hectare)	7.3	- 19.5	- 26.8	- 45.3
2.	ΔNet revenue (%)	- 5.5	- 11.3	-20.9	- 34.5

Source: Ariel et al., 2008

Institutions that promote agricultural development, environmental resource management, disaster, health, transport and water management require coordinated action plans to counter the expected climate change effects. The rural poor are the most vulnerable due to their low purchasing power and level of education. Therefore the government has the duty to seek for the means of mitigation of their sufferings through sustainable adaptation strategies. The most appropriate adaptation strategies should focus on the use of local resources to enhance the diversification of livelihoods.

Awareness and Adaptation at Household Level

The direct impacts of climate on farming as a livelihood activity has necessitated adaptations in farm practices and cropping patterns in response to changed weather conditions. These adaptations at farm family level include:

- Shifting crop mix to more drought tolerant and short-season varieties.
- Reducing the area planted initially, then increasing it gradually, depending on the nature of the season.
- Staggering early or late planting dates.
- Increasing plant spacing
- Maximizing the use of clayey soils where these are available due to their high moisture retention capacity.
- Mulching of the soil in order to reduce evaporation and to increase the soil moisture reserve.
- Use of organic manure in order to increase the soil moisture retention capacity.
- Use of rainwater harvesting techniques in the drier agro-ecological zones.
- Undertaking indigenous religious ceremonies to appeal to nature to bring more rain and to avert droughts.
- Vulnerable households adapt to changing conditions due to climate change by switching to wild sources of food. These sources are important for the poor and the landless in times of famine and insecurity or conflict when food supply mechanism are disrupted and local or displaced people have limited access to food. Wild foods are important in complementing staple foods. Wild sources of food therefore offer income opportunities to the rural poor.
- Temporary migration is another adjustment strategy adopted by rural households. Tyson *et al.* (2002) find a significant relationship between climate patterns and temporary migration. These migrations in Cameroon are generally southwards from the drier agro-ecological zones. In West Cameroon these are from highlands to flood plain topographic sites in river basins. There is also the movement of labour from

agriculture into more productive sectors as well as rural-urban migration (D Haan, 1999) Migration offers the potential to enhance household ability to withstand shocks and stress through alternative opportunities.

- In the Sudan and sahel savannah ecological climates of the country pastoral people migrate in response to annual and seasonal rainfall variations. Desanker (2002) suggests that pastoral systems are intrinsically able to adapt to fluctuating and extreme climates provided they have sufficient scope for movement, and if other necessary elements sustaining the system remain in place.
- Diversification of income earning by rural households is another adaptation strategy. Seasonal effects and climatic uncertainty that characterize the agricultural sector oblige farmers to diversify sources of income and employment opportunities (Downing *et al.,* 1997).
- Replanting and crop sequencing are adaptations in the savannah ecological zones of Cameroon. Adjusting the cropping sequence, including changing the timing of sowing and planting to take advantage of the changing duration of thc growing scason and associated heat and moisture levels alter the length of the growing season to better adjust to the changed environment.

The above indigenous adaptations to climate change by farmers in Cameroon raise an important policy question, that is, whether agricultural adaptation in the face of climate change can be expected to occur automatically or whether government has a role in intervening. At least some adaptation takes place automatically. However, the government can have an important role in promoting certain adaptation strategies and technologies:

- developing breeding programmes for drought resistant varieties.
- developing more efficient irrigation systems and enabling the technologies to be affordable to farmers.
- developing more efficient rain water harvesting techniques and extending the technologies to farmers.
- investing in water storage and distribution structures and an on-farm delivery systems for farmers.
- developing seed banks and food storage facilities to enhance availability in drought or flood years.

- promoting crop and livestock diversification in order to enhance risk management.
- breeding for early maturing crop varieties and high yielding varieties.
- promote the breeding and cultivation of low input varieties.

Mortimore (1989) identifies the ways farmers in the sudano-sahelian agro-ecological zone adapt to drought and hunger. The strategies to mitigate drought include crop mixtures, re-plantings and manure inputs. For hunger management, people resort to eating leafy foods harvested from the wild. Degrading biodiversity as a result of climate change may therefore lead to reduced food security. This is because changing conditions cannot always be buffered by genetic or species diversity. The risk of famine will therefore increase with climate change, particularly when agricultural systems and livelihood options rely on few or only one species and institutions cannot provide support.

There is a need for government assistance to the rural populace in the integration of community adaptation strategies into policy and in the development of appropriate technologies. The development of agroforestry systems for each agro-ecological zone is a possibility for combining biodiversity conservation and climate change adaptation. Agroforestry systems are more resilient against climate change than tree-less fields, as the trees reduce wind speeds and therefore evaporation, and trap water and nutrients. They harbour higher species diversity, including, for example, predators that check insect pests. Breeding programmes on government research stations must enhance genetic crop diversity. High agro-biodiversity allows for local adaptation processes. Supporting farmers in maintaining their locally adapted native varieties is an important insurance against future increase of temperature variation and the occurrence of new pests and diseases. It is an even better safeguard than genetically modified crops which often have a reduced genetic variability.

Sustainable Financing of Adaptative Strategies

Cameroon like most sub-Saharan African countries is faced with additional challenges:

- The prevailing socio-economic situation means that the design and promotion of adaptation strategies will involve higher opportunity costs. This is because the resources involved could often also be used to promote other aspects of rural development.
- The problems of poverty exacerbate the competition for scarce resources and mean that the promotion of agricultural adaptations to climate change has a relatively low priority.
- Faced with these problems and that of good governance, Cameroon requires support from the industrialized countries. Endowment funds, from which only the interest on the deposited capital is used, are necessary. Such funding requires conditions of subsidiarity and good governance.
- Transparent administration and supervisory structure in accordance with civil society principles;
- Criteria for awarding grants from funds on the basis of priorities;
- Functioning management structures for the climate change adaptation strategies under consideration;
- Existence of management, operating and financial plans;
- Clearly formulate target indicators and work indicators for the design, promotion and management of adaptation technologies and structures.

Climate Change, Governance and Agricultural Development

Cameroon urgently needs a National Adaptation Programme of Action (NAPA) and a National Communication on climate change which will take care of medium and long-term measures. Both programmes are endorsed by the United Nations Framework Convention on Climate Change (UNFCCC). However, the country does not lack programmes and institutions dealing with climate change:

- *The Ministry of Forest and the Environment.*

This ministry in collaboration with the United Nations Environment Programme (UNEP) has set up a climate change unit charged with environmental assessment.

- *The Ministry of Agriculture and Rural Development:*

The ministry does not have a defined climate change impact assessment programme and adaptation strategy in their agricultural extension programme. Adaptation is however, undertaken only by extension workers who perceive climate change. The perception of climate change appears to hinge on farmer experience and the availability of extension advice specifically related on climate change. But while the policy options for promoting an increased awareness of climate change are more limited, the perception of climate change by officials in this ministry is already high. In the drier Sudano-Sahelen zone of the country famine imposed by droughts have obliged this ministry to establish grain banks that alleviate suffering in times of crisis.

- *The Ministry if Scientific research:*

This ministry undertakes agronomic research in the different agro-ecological zones. Crop breeding programmes have a focus on drought resistance, early maturing varieties, resistance to windstorms and cultural practices. However, there is no formal policy geared at climate change adaptation. Adaptation programmes to climate change must take into account crop selection. Treating crop choices as exogenous will seriously overestimate the changes from global warming. For example, agronomic studies or empirical studies that use weather as a proxy must be careful not to assume crop choices are exogenous. Farmers will probably change crops in response to a new climate rather than repeatedly grow crops that historically were successful but now fail. As a result, farmers will match future crops to future climates. There is therefore an important role for agronomic research to develop new varieties more suited for higher temperatures.

- *Ministry of Livestock and Animal Industries:*

The ministry has no climate change adaptation extension programmes. However, endogenous developments in the grassland ecosystems of North Cameroon following the droughts of the 1970s and 1980s indicate that herders are switching species, diversifying their portfolio of animals, and moving away from cattle and towards sheep. Small farmers are expected to make these changes without much change in expected net income. Policy makers must be careful to encourage private adaptation during this period of change. There may be nothing that can be done to sustain large cattle herders that depend on current climate. Providing subsidies and other enticements for such operations to continue once the climate changes would only compound the problem. Instead, the government should encourage farmers to change the composition of animals on their farms as needed. There is a need to share indigenous knowledge on how livestock owners have coped with higher temperatures. The government should anticipate that farmers will make changes on their lands and do whatever is needed to facilitate these changes. Some indigenous adaptations include: herd supplementation, herd movement following rainfall opportunistic events and edaphic floristic communities, livestock diversification, culling animals and reduction of herd / farm sizes.

In spite of the vantage point of climate change in the international politics and media the political will to counter climate change in Cameroon is only coming up now in implicated ministries. For measures to be successful futures climate change condition adaptations must be an intrinsic part of international and national legislation as well as government and administration. Current institutions that are supposed to tackle this challenge posed by climate change do not have reliable programmes. There is need for a coordinated programme of action. Sustainable adaptation will hardly be achieved without improved human and institutional capacity building. There is the need to develop partnerships with global institutions that possess resources and lessons for climate adaptation strategies as stated under the goal number 8 of the Millennium Development Goals.

The development of a policy on climate change mitigation and adaptations to global warming is consistent with the Cameroon Environmental Management Plan of 1994 which governs the management of the environment and natural resources (Figure 2). The planning framework proposes an integration approach. The present management of natural resources and the environment is sectorial with many weaknesses. The climate change management plan should have the following key elements as a central focus:

- *Climate change science*: Monitoring of temperature and precipitation
- *Climate change vulnerability*: Monitoring the sensitivity of natural and human ecosystems to climate change (Impacts on biodiversity, food security and water).
- *Climate change adaptation*: Developing technologies to mitigate climate change, buffer the influence of climate change as well as supporting local people in adapting to it and in minimizing risks.

Education and awareness to climate change impacts: Facilitating the above directions by placing emphasis on information, advice and advocacy together with opportunities for meaningful grassroots participation in decision-making.

These elements are central to an effective and sustainable climate change management policy.

The second key element is the development of site-specific mitigation and adaptation strategies within each agro-ecological region and for each farming system. The complexity of the problem requires specific adaptation strategies developed using a multi-stakeholder approach involving the central government, related ministries, regional governments, local governments, and other related local, national and international stakeholders.

The third key element involves investments in the above key elements, that is, developing and instituting a national structure for climate change management, adaptation technology, development and diffusion of adaptation technologies and river basin modeling.

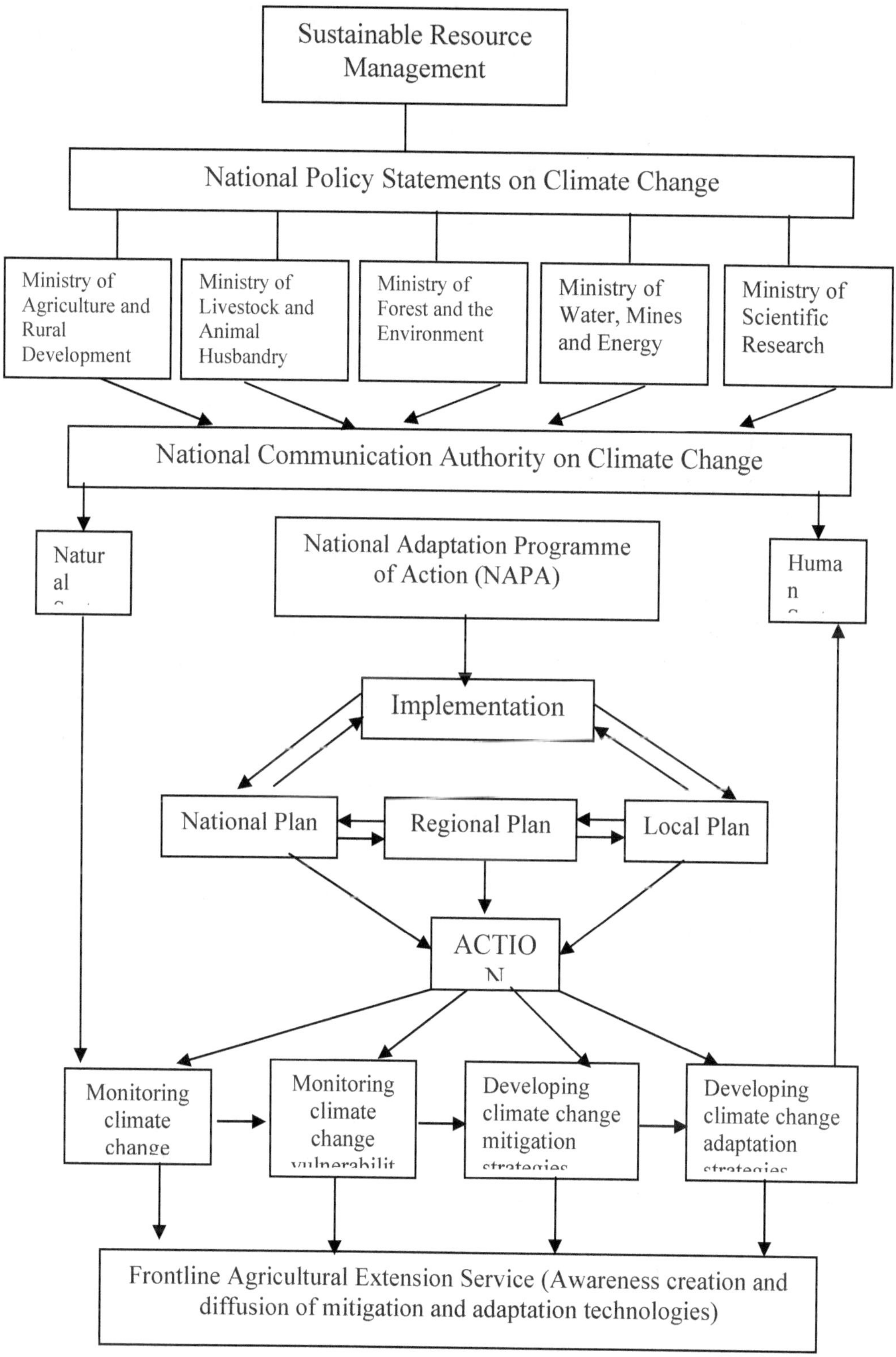
Sustainable Resource Management
National Policy Statements on Climate Change
Ministry of Agriculture and Rural Development
Ministry of Livestock and Animal Husbandry
Ministry of Forest and the Environment
Ministry of Water, Mines and Energy
Ministry of Scientific Research
National Communication Authority on Climate Change
Natur al
National Adaptation Programme of Action (NAPA)
Huma n
Implementation
National Plan
Regional Plan
Local Plan
ACTIO N
Monitoring climate change
Monitoring climate change
Developing climate change mitigation
Developing climate change adaptation
Frontline Agricultural Extension Service (Awareness creation and diffusion of mitigation and adaptation technologies)

Figure 2: *Planning framework for the development of climate change Policy in Cameroon.*

Agricultural Development and Extension Measures

Cameroon needs to institute a Department of Environment Assessment in the Ministry of Environment and Forests as well as an Inter-Ministerial Committee on climate change and a National Coordinating Committee on Climate Change. The implicated ministries include the Ministry of Agriculture and Rural Development, Ministry of Livestock and Animal Industries, Ministries of Research and the Ministry of Environment and Forests. Their main policy implications for the adaptation of agriculture to climate change include the following:

- Mitigation of climate change impacts on crop water requirements: The development of appropriate and efficient, simple farmer-operated small-scale irrigation technologies will allow farmers to adapt much better. This will require an analysis of the costs and benefits. Clearly, irrigation provides higher yields that may justify the investment. By careful design of an incentive system for both technology adoption and location, selection for certain crops, the government can enhance food production and reduce water use even under severe climatic conditions.
- Endogenous adaptation strategies and barriers to adaptation: There is evidence that agricultural adaptation to climate change takes place automatically. Farmers seek to adapt their cropping practices and farming systems to changing climatic circumstances. The government can have an important role in testing, validating and promoting some of those adaptation strategies and in removing barriers to adaptation.
- Expansion of crop choices: there has in recent years been a heightened interest in agroforestry and agro-silvopastoral practices, integration of animals in cropping systems and adjustments of crop sets and cultural practices to changing climate conditions. The government breeding and farming research programmes must enhance these adaptations.

- River-basin hydrology and modeling: Government policy options in this direction include both the spatial and inter-temporal variability in stream flow. Spatial variability is great between the sudano-sahelian regions of North Cameroon and the moist humid regions of the south. Second, both the low and high ranges of inter-temporal flow will widen. This will result in frequent floods and droughts. Bracing for these likely events by introducing adaptation measures earlier rather than later should become a priority of the government and development agencies. Cameroon shares the Chad basin, Niger basin, and Congo basin with other countries. Coordination among these riparian states sharing the same basin should address both water management and dam operations to allow smoothing events when they strike.

Conclusions and Recommendations

Climate change and vulnerability constitute multiple threats to sub-Saharan African countries. The key risks hover on the agricultural systems, and the riverine and dam-neighbouring communities. It is unfortunate that Cameroon like many Sub-Saharan countries has very few institutions dealing with climate change adaptations. In spite of the vantage position of climate change in international politics and the media, the political will to counter climate change in Cameroon is yet to fully emerge. Like many parts of Africa, the large share of its economy depends on climate sensitive sectors like rainfed agriculture. Despite the emergence of indigenous adaptation strategies, Cameroon and most of Africa lags behind the rest of the world in adopting irrigation, capital and high yielding varieties. Policy makers may want to pay special attention to some successful adaptation strategies which can help farmers adapt to drier or hotter conditions. Through research and out reach the government can encourage the development and the use of crop varieties with more tolerance to hot and dry conditions. Diversification of crops, crop varieties and animal breeds, and cultural practices based on comparative agroclimatic and economic advantage remain key adaptation measures. Promoting irrigation and mixed crop-livestock systems can help alleviate the likely effects of climate change. Where water is available switching from rainfed to irrigated agriculture will increase not only average net revenue per hectare, but also the resilience of agriculture to climate change. The government needs to make public investments in infrastructure and canals for water storage and delivery where appropriate. Policy makers may consider supporting adaptation interventions for climate change where appropriate.

The perception of climate change hinges on farmers' experience and the availability of extension advice specifically related to climate change is already high, therefore, enhanced farmer education can facilitate adaptation. This implies the provision of an effective extension system and good farm-to-market roads. Improved transport linkages to markets can enhance adaptation to climate change. Better farm-to-market roads may allow farmers to switch from subsistence farming to cash crops, or facilitate the exchange of ideas, and innovation diffusion through interaction with markets and market demands. Finally, sustainable adaptation can only be achieved if these measures are accompanied by institutional and human capacity building, inter-sectorial and ministerial collaboration, developing partnerships with related global institutions, and good governance. The adaptation of agriculture to climate change must be an intrinsic part of international and national legislation as well as government and administration.

References

Adelekan, I. (2000) A survey of rainstorm weather hazards in Southern Nigeria. *Environmentalist Vol. 20,* No. 1, p. 33-39.

Aliyu, S. (2008) An appraisal of climate change risks and institutional adaptation strategies in Kano State. *Book of Proceedings: Geography and Millennium Development Goals – translating Vision into Reality in Nigeria.* University of Calabar. p. 121-125.

Ariel, D.; Hassan, R.; Mendelsohn, R.; benhin, J. (2008) Climate change and agriculture in Africa. Earthscan, London. 189p.

Awosika, L.; French, C.; Nicholas, R.; Ibe, C. (1992) The impact of sea level rise on the coastline of Nigeria. In: O'Callahan J (ed) *Global climate and the rising challenges of the sea.*Proceedings of the IPCC Workshop at Mongarine Island, Venezuela, March 1992. p. 9-13.

Ayonghe, S. (2001) A quantitative evaluation of global warming and precipitation in Cameroon from 1930 to 1995 and projections to 2060: effects on the environment and water sources. In: C.M. Lambi and E.B.Eze (eds) *Readings in Geography;* Unique Perinters, Bamenda. p. 142-155

Buadi, P. and Ahmed, A. (2006) The impacts of wind related hazards in Southern Cameroon. *Journal of Physical Sciences Vol. 3, No. 2.* p. 82-88.

Carter, R. and Alkali, A. (1996) Shallow groundwater in the northeastern arid zone of Nigeria. *Quarterly Journal of Engineering Geology, vol. 29,* p. 341-355

De Haan, A. (1999) Livelihood and poverty: The role of migration – a critical review of the migration literature. *Journal of Development Studies, Vol. 36, No.2,* p.1-47.

Downing, T.; Ringius, L.; Hulme, M.; and Waughray, D. (1997) Adapting to climate change in Africa. *Mitigation and Adaption Strategies for Global Change, Vol. 2. No. 1*, p. 19-44.

F.A.O. (2007) Climate change and food security. (http/www.fao.org/nr/index-en.htm

Fondo, S. (2001) The socio-economic impacts of potential climate change on the Cameroon coastline. In. C.M. Lambi (ed.) *Environmental Issues: problems and prospects.* Unique Printers, Bamenda, p. 159-173.

IPCC (2001) The report of working group 1 of the Intergovernmental Panel on climate Change Survey for Policymakers. IPCC Report.

MacGregor, J.; Reid, H.; Sahlen, L. (2006) An agenda for change: Quantifying climate change impacts on natural resource-based economies. *Sustainable Development Option Papers.* October, 11057 HED.

Malun, C. (2006) Socio-economic impacts of climate change in Cuba. *Journal of Environmental Science, Vol. 8, No.2.* p. 44-60

NEST (2004) Climate change in Nigeria: a communication guide for reporters and educators. NEST, Ibadan.

Nyelong, P. (2004) Global warming and global waters. *Journal of Energy and Environment, No. 17, No. 1.* p. 79-90.

Odinga, S. (2004) Evaluation of the economic cost of climate change in Mozambique. *Environmental Impact Analysis, Vol. 12, No. 1,* p. 45-52.

Odjugo, P. (2000) The impacts of global warming on climate change and extreme weather conditions: Global and Regional evidences. *Asia Pacific Journal on Environment and Development, Vol. 7,* p. 53-70.

Osman-Elasha, B. (2007) Africa's vulnerability. *Teimpo, A Bulletin of Climate and Development, Issue 63,* April, 2007.

Reid, H; MacGregor, J.; Sahlen, L. and Stage, J. (2007) Countering the cost of climate change in Namibia. *Sustainable Development Option Papers,* December, 17026 HED.

Tyson, P.; Lee-Thorp, J.; Holmgren, K.; and Thackeray, J. (2002) Changing gradients of climate change in Southern Africa during the past millennium: implications for population movements. *Climate Change, Vol. 52,* p. 129-135.

Velarde, S.; Malhi, Y.; Moran, d.; Wright, J. and Hussain, S. (2005) Valuing the impacts of climate change on protected areas in Africa. *Ecological Economics, Vol. 53, No. 1,* p. 21-33.

www.ingramcontent.com/pod-product-compliance
Lightning Source LLC
LaVergne TN
LVHW052035200926
848481LV00005B/1684